Aid and Development

This book provides an overview of what aid is, how it has changed over time and how it is practiced, as well as debates about whether aid works, for whom and what its future might be.

The text shows how 'aid' is a contested and fluid concept that involves a wide and changing variety of policies, actors and impacts. It equips the reader with an understanding of what aid is, where it comes from and where it goes, how it is delivered and what its impacts are, and whether shortcomings are a result of a fundamental problem with aid, or merely the result of bad practices. It explores the changing political ideologies and conceptions of development that continually reshape how aid is defined, implemented and assessed, and how, despite a global commitment to the Sustainable Development Goals, we are at a point where the very notion of aid is being questioned and its future is uncertain. Each chapter includes case studies, chapter summaries, discussions, weblinks and further reading, to help strengthen the reader's understanding.

Aid and Development provides an important resource for students, development workers and policy makers seeking an understanding of how aid works.

John Overton is a geographer who has worked on development issues in Africa, Southeast Asia and the Pacific Islands for some forty years. He has particular interests in both rural transformations and the changing conceptions and practices of international aid. He is currently Professor of Development Studies at Victoria University of Wellington, Aotearoa/New Zealand.

Warwick E. Murray is Professor of Human Geography and Development Studies at Victoria University of Wellington, Aotearoa/New Zealand. He has held university positions in the UK and Fiji. He is currently President of the New Zealand Geographical Society and past-President of the Council for Latin American Studies of Asia and Oceania. He has served as an editor of various journals including the *Journal of Rural Studies* and *Geography Compass*, and is Editor-in-Chief of *Asia Pacific Viewpoint*.

Routledge Perspectives on Development

Series Editor: Professor Tony Binns, *University of Otago*

Since it was established in 2000, the same year as the Millennium Development Goals were set by the United Nations, the Routledge Perspectives on Development series has become the pre-eminent international textbook series on key development issues. Written by leading authors in their fields, the books have been popular with academics and students working in disciplines such as anthropology, economics, geography, international relations, politics and sociology. The series has also proved to be of particular interest to those working in interdisciplinary fields, such as area studies (African, Asian and Latin American studies), development studies, environmental studies, peace and conflict studies, rural and urban studies, travel and tourism.

If you would like to submit a book proposal for the series, please contact the Series Editor, Tony Binns, on: jab@geography.otago.ac.nz

Information and Communication Technology for Development (ICT4D)
Richard Heeks

Media and Development
Richard Vokes

Education and Development
Simon McGrath

Postcolonialism, Decoloniality and Development, 2nd edition
Cheryl McEwan

South-South Development
Peter Kragelund

Gender and Development, 3rd edition
Janet Momsen

Aid and Development
John Overton and Warwick E. Murray

For more information about this series, please visit: www.routledge.com/series/SE0684

Aid and Development

John Overton and Warwick E. Murray

LONDON AND NEW YORK

First published 2021
by Routledge
2 Park Square, Milton Park, Abingdon, Oxon OX14 4RN

and by Routledge
52 Vanderbilt Avenue, New York, NY 10017

Routledge is an imprint of the Taylor & Francis Group, an informa business

British Library Cataloguing-in-Publication Data

A catalogue record for this book is available from the British Library

Library of Congress Cataloging-in-Publication Data
Names: Overton, John (John D.), author. | Murray, Warwick E., author.
Title: Aid and development / John Overton and Warwick E. Murray.
Description: Abingdon, Oxon ; New York, NY : Routledge, 2021. |
Series: Routledge perspectives on development |
Includes bibliographical references and index.
Subjects: LCSH: Economic assistance. | Economic assistance—Evaluation. |
Economic assistance—Case studies.
Classification: LCC HC60 .O96 2021 (print) | LCC HC60 (ebook) | DDC 338.91—dc23
LC record available at https://lccn.loc.gov/2020016215
LC ebook record available at https://lccn.loc.gov/2020016216

ISBN: 978-0-367-41483-2 (hbk)
ISBN: 978-0-367-41484-9 (pbk)
ISBN: 978-0-367-81475-5 (ebk)

Typeset in Times New Roman
by codeMantra

Contents

Figures

Tables

Boxes

Acknowledgements

This book had its genesis in a number of university courses we teach. Students over a number of years have produced assignments and asked questions that have made us think more critically about aid. Student research for PhD and masters' theses by Avataeao Junior Ulu, Faka'iloatonga Taumoefolau, Felicia Talagi, Helen Mountford, Finbar Kiddle, Thomas McDowall and others has deepened our understanding of how aid relationships work in practice in the Pacific region. As a colleague and 'pracademic' *par excellence*, Gerard Prinsen has opened our eyes to the intricacies and importance of aid modalities, financial management systems and the like. And many conversations with Gerard and other colleagues at the Paekakariki Institute of Social Sciences, DevNet and NZADDS – Nicki Wrighton, Regina Scheyvens, Glenn Banks, Emma Mawdsley, Terrence Wood, Jo Spratt, Luke Kiddle, Andrew McGregor and many others – have helped shape and question the way we see aid. And for this book, we wish to acknowledge the very helpful comments of Tony Binns and anonymous reviewers which helped to sharpen our material. We thank all of the above for contributing to our education in aid. However, given the range of views about aid, its worth and its future – and long may such debates continue – we alone take responsibility for the views expressed in this book.

Abbreviations

ACP	African, Caribbean and Pacific countries
ADB/AsDB	Asian Development Bank
AGCI	*Agencia de Cooperación Internacional de Chile* (Chile)
ART	Antiretroviral Treatment (for HIV/AIDS)
BRICS	Brazil, Russia, India, China, and South Africa
CARE	Cooperative for Assistance and Relief Everywhere (formerly Cooperative for American Remittances to Europe)
CDC	Commonwealth Development Corporation
CSSF	Conflict, Stability and Security Fund (UK)
DAC	Development Assistance Committee (of the OECD)
DCD	Development Co-operation Directorate (of the OECD)
DFID	Department for International Development (UK)
DIB	Development Impact Bond
EBA	Enabling the Business of Agriculture (World Bank)
ECLAC	Economic Commission for Latin America and the Caribbean
ERP	European Recovery Program (Marshall Plan)
EU	European Union
FAO	Food and Agriculture Organisation
FDI	Foreign Direct Investment
GBS	General Budget Support
GDP	Gross Domestic Product
GEF	Global Environment Facility
GFC	Global Financial Crisis
GNI	Gross National Income
GNP	Gross National Product
HDI	Human Development Index
IATI	International Aid Transparency Initiative

IBRD	International Bank for Reconstruction and Development (World Bank)
ICAI	Independent Commission for Aid Impact (UK)
IDA	International Development Association (World Bank)
IFC	International Finance Corporation (World Bank)
IMF	International Monetary Fund
IQC	Indefinite Quality Contract
ISI	Import Substitution Industrialisation
ISIS	Islamic State
LDC	Least Developed Country
MDGs	Millennium Development Goals
MFAT	Ministry of Foreign Affairs and Trade (New Zealand)
MICEI	Magrabi ICO Cameroon Eye Institute
MOFA	Ministry of Foreign Affairs (Japan)
MOFAIC	Ministry of Foreign Affairs and International Cooperation (United Arab Emirates)
MOFCOM	Ministry of Commerce (China)
MSF	*Médecins sans Frontières*
NGO	Non-Government Organisation
NSP	National Solidarity Programme (Afghanistan)
OCHA	Office for the Coordination of Humanitarian Affairs (UN)
ODA	Official Development Assistance
ODI	Overseas Development Institute (UK)
OECD	Organisation for Economic Cooperation and Development
OPIC	Overseas Private Investment Corporation (USA)
PEFA	Public Expenditure and Financial Accountability
PIDG	Private Infrastructure Development Group
PPP	Public-Private Partnership
PRC	People's Republic of China
PRSP	Poverty Reduction Strategy Paper
PSF	Peace and Stabilisation Fund (Denmark)
RoC	Republic of China (Taiwan/ Chinese Taipei)
RSE	Recognised Seasonal Employer (New Zealand)
SAPs	Structural Adjustment Programs
SDGs	Sustainable Development Goals
SWAp	Sector-Wide Approach
UN	United Nations
UNAIDS	United Nations Programme on HIV/AIDS
UNDP	United Nations Development Program
UNEP	United Nations Environment Program
UNESCO	United Nations Educational, Scientific and Cultural Organisation

UNHCR	United Nations High Commission for Refugees
UNICEF	United Nations Children's Fund
USAID	United States Agency for International Development
VfM	Value for Money
VGF	Viability Gap Funding
WFP	World Food Program
WTO	World Trade Organization

1 Aid

An introduction

Learning objectives

This chapter will help readers to:

- Understand some of the varying definitions of aid
- Appreciate in a broad sense the historic roots of aid and its present geography
- Understand the role of aid in the global economy and how this is changing
- Debate the justifications for and motivations of aid from different perspectives
- Understand in outline form the arguments of critics and supporters of aid

Introduction

In the Global North, the term 'aid' often conjures up images of human suffering being met by assistance from outside. Such assistance has become an integral part of the way we conceive and practice development. We see aid as a means to alleviate suffering but also to promote development and self-reliance. Such views have been reinforced over the years by well-meaning public campaigns which have encouraged the public to both dip into their own pockets to contribute to relief efforts and pressure their governments to increase their aid budgets.

However, as we will see, there have been many critics of aid: those who have suggested that aid does not achieve what it says it aims to; that it merely acts as another weapon for the rich and powerful to exploit the poor and vulnerable; or that it actually distorts the economy and makes poverty worse. These debates continue to rage.

Some commentators have also suggested that the age of aid is coming to an end – we are entering an historical 'post-aid' period where new actors, new ways of operating and challenges to the old simplistic world order of rich aid donors and poor recipients, are ushering in fundamentally different sets of economic and political relationships and power structures (Mawdsley *et al.* 2014; Mawdsley 2018; Janus *et al.* 2015; Gulrajani and Faure 2019).

Yet we contend that forms of aid from some countries to others are still significant features of the global economy. Furthermore, there are signs that aid will continue to be used to support global ambitions to alleviate poverty, tackle the effects of climate change, lessen inequalities and promote economic growth. The Sustainable Development Goals of 2016–30, in particular, set some lofty objectives and specific targets to address major global challenges for the next decade (see Box 1.1) – and if these are to be pursued seriously, they will require significant funding from both public and private sources. Aid, in its many forms, will remain important for global development efforts for years to come. Therefore, there is a need for us to understand aid in its many forms and how it has changed over time, to see how it is delivered, to question its effectiveness and to identify lessons for the future.

Box 1.1 The Sustainable Development Goals 2016–30

Following the Millennium Development Goals (MDGs) of 2000–15, the United Nations launched the Sustainable Development Goals (SDGs) in 2016 ostensibly to set the development agenda for the next 15 years. Unlike the MDGs, the SDGs were agreed on following a long period of consultation and they attempted to incorporate key environmental concerns and goals and alongside more poverty-related objectives. In addition, whereas the MDGs tended to focus on the developing world, the SDGs were more inclusive, recognising that development and environmental concerns affected every country and that poverty and environmental degradation were not confined to so-called 'poor' countries. The slogan 'leave no-one behind' represented an ambitious and all-encompassing mission to transform the way we think about and pursue development.

Seventeen Goals were set. These encompassed concerns embedded in the MDGs regarding poverty, hunger, health, education and gender equality. To these were added goals relating to environmental sustainability: clean water, clean energy, the health of oceans and land, responsible consumption and production and climate action. Furthermore, there are goals relating to the economic sphere: industry and infrastructure 'decent' work and economic growth. Obscuring the debates whether such goals are consistent, mutually compatible or attainable, there is also an interesting social objective in the form of 'reduced inequalities'. The 17 Goals are supported by 169 targets and a large list of 232 targets.

The Goals represent a remarkable degree of global consensus regarding the crucial issues and priorities facing the planet. Environmental and development-related processes and outcomes are, of course, closely inter-related, as are economies, societies and environments world-wide. The SDGs do well to recognise this holistic and inter-connected framework. Yet the SDGs will require a great deal of commitment and adaptation so that local needs and priorities can shape on-the-ground action and change.

The SDGs will also require a huge investment of resources at the global scale if they are to be achieved. One estimate from *The Economist* is that the SDGs will cost between $US2–3 trillion per year or 4 per cent of global gross domestic product (GDP) (*The Economist* 2015). There is talk of both private and public capital being needed in very large amounts. Private investment and philanthropy will need to be complemented by public commitments – much of it in the form of Official Development Assistance (ODA). And, given the huge scale required, ODA would have to be increased to well beyond the 0.7 per cent target set, but not achieved, by all but a few donor countries. Whereas the MDGs were associated in the early 2000s with apparent public support for increased aid and things such as debt relief and poverty alleviation, the global political environment in the later 2010s appears to be rather less supportive of substantial increases in public funding for aid.

Therefore, the SDGs require us to look again critically at how aid works, or not, and how it might be used to address this new 2030 development agenda.

In this book we aim to better understand what aid is, and has been, where it comes from and where it goes, how it is dispersed and what its impacts are. We need to question whether any observed shortcomings are a result of a fundamental problem with aid, or merely the result of bad practices. We need to look at the motives and policies of donors as much as we do the conditions and efforts of recipients. We provide students of development and those who work, or intend to work, in the development and aid sector with a broad picture of the present aid 'landscape', together with some key concepts and methods, and an overview of debates concerning the impacts and possible futures of aid. We adopt a broad definition of 'aid' and appreciate its complexities and dynamism, with emerging new actors and modes of operation, though we continue to focus on dominant framings of aid, key agencies and mainstream ways of operating, as seen in the OECD-defined definitions and measurements of ODA.

In this first chapter we briefly outline the position of aid in the global economy and the motivations for giving and receiving it before examining some of the key criticisms and debates. The core chapters of the book then seek to address some key questions:

- What is aid? How is aid defined and measured in various ways and how might various forms of 'assistance' or 'co-operation' be considered aid, or not? (Chapter 2)
- What is the geography of aid in terms of volumes and flows? Who are the major donors and recipients and what are the key aid agencies? (Chapter 3)
- How has aid changed over time? How have the principles, objectives and methods of aid delivery evolved through various historical 'regimes' of aid? (Chapter 4)
- How is aid delivered? In what forms does aid appear, what are the various scales of operation and how do these different aid 'modalities' involve different actors? What new forms of aid delivery are emerging at present? (Chapter 5)
- Does aid work? How can we start to understand the effects of aid on economic systems, governance, welfare and social structures – and what debates exist regarding the impacts of aid? (Chapter 6)
- What have we learned about effective aid and what is the future of aid? (Chapter 7)

Aid and development

Before proceeding further, we need to pause and consider what we mean by 'aid' and 'development'. Firstly, although we will examine the definition of 'aid' in some depth in Chapter 2, here we can suggest that aid involves some broad idea of 'help' or 'assistance' from one party to another. It may – and certainly in practice almost always does – involve two-way flows (whether these involve financial resources, political favours, technical advice, trade agreements or movement of people) but 'aid' should imply that at least the initial flow is from a donor to a recipient and that this involves some notion of assistance. In this book, we focus on aid which has some sort of development or humanitarian objective – such as economic growth, improved welfare or disaster relief – and most commonly we resort to the widely used definition of development aid as ODA, and focus on aid which is delivered and received by state and civil society agencies

Table 1.1 ***Some key terms used in this book***

'**Aid**': a general term referring to assistance that can come in many forms (financial, technical, material or non-material) and which may involve two-way flows of costs and benefits
'**Development aid**': resources, whether financial or technical, given by various donor agencies to assist poorer countries and communities undertake economic, social and political development. It focuses on building the long-term capacity of economies and institutions to undertake their own development in future.
'**Humanitarian aid**': a short-term response to provide needed food, shelter, medical supplies and other assistance following natural disasters or civil disturbances and violence.
'**Official Development Assistance**' (ODA): 'government aid designed to promote the economic development and welfare of developing countries. Loans and credits for military purposes are excluded … Aid includes grants, 'soft' loans (where the grant element is at least 25% of the total) and the provision of technical assistance' (OECD 2019b).
'**Development co-operation**': joint actions and contributions from different parties (usually governments) to pursue strategies and activities that address defined development goals and which may accrue benefits to all partners.

(rather than, say, aid given within families), although we acknowledge that other forms of aid may exist (Table 1.1).

We should also note that the very concept of aid has been challenged. Writing in *The Development Dictionary,* Marianne Gronemeyer not only points to the 'perversion of the idea of help' (2010: 57), through military aid or food aid (allowing the spread of global corporations selling seed grain), but also sees aid as a form of 'elegant power': 'elegant power does not force, it does not resort to the cudgel or to chains; it helps' (2010: 55). To her, development aid is not innocent, welcome or unconditional but 'aid' is a linguistic device used to pursue the self-serving objectives of the 'giver' and impose power over the 'needy'.

The task of defining 'development' is even more complicated than agreeing on meanings of 'aid'. Most conventional understandings of development suggest improvements in material standards of living, commonly measured by national income (e.g. Gross National Product (GNP)/Gross National Income (GNI) per capita). These approaches highlight the objectives of increasing economic activity, employment, productivity and growth and they typically use aggregate approaches, assuming that overall growth and improvement will filter down to all in an economy. We see this approach prominently in discussions of aid and its effects: how (or if) aid promotes economic growth or helps create jobs, for example. Yet we also know that 'development' can – and should – address more qualitative and less overtly material aspects such as human rights, well-being and happiness, individual freedom, social justice, sustainability, dignity and security. Furthermore, development goes deeper than aggregate processes or measurements: social inequality, political power imbalances and economic systems mean that there are both winners and losers in development and our focus arguably should be on the poor, marginalised and dispossessed rather than the economy as a whole. Indeed, it is often these aspects, particularly recognition of human suffering at an individual or community level, that is used to portray 'need' and justify the giving of aid. In this book, we adopt this broad approach to development, seeing it as a complex and contested notion of 'improvement' or 'good change' (Chambers 2004), though conceding that processes of development can lead to both positive and negative outcomes for different parties.

Therefore, we can see aid and development as difficult concepts to define precisely. They are also fluid. In this book, we suggest that aid has changed significantly over time, often in response to different contemporary understandings of what development is or should be. Certain understandings and theories have become embedded as broadly agreed and dominant approaches in the delivery of aid adopted by key agencies at different times in recent history. These are each associated with different political ideologies, motivations, power relationships, methods of operating and institutions. They are 'regimes', or collections of ideas, institutions and practice, that exist for a period of time before giving way to another dominant approach and they are similar in this sense to paradigms. In Chapter 4 we will suggest that four main regimes of aid can be identified since 1945: modernisation, neoliberalism, neostructuralism and retroliberalism. These are important to understand for they have shaped the aid world in quite marked and varied ways over time. Elements of each have been dominant at different times, though more than one may be evident at any one moment in time.

Aid in the global system

International development assistance has been one of a number of major flows of financial capital across the globe. It sits alongside foreign direct investment (FDI), private remittances and international trade as one of the key elements of the global economy. As with other forms, the flows of aid are not all one-way from rich to poor countries, for there are return movements of interest on and repayment of assistance loans and (even though such loans are generally given on a concessional basis) the net flow over time may actually be from poor to rich. Yet aid is the one flow that is predicated on a deliberate intervention in the global economy to change the flow of resources so that one group of countries and societies ('developing'[1] countries) can benefit from assistance from another ('developed' countries).

In relative terms, aid[2] is not the major source of financial flows to developing countries. It is outstripped by both private remittances (personal savings or money people send to family members and others 'at home') and FDI. In 2015 aid, or as it is known formally ODA, accounted for just 16.5 per cent ($US179.5 billion) of total resource receipts to developing countries (down from 18 per cent in 2000) (OECD 2017 – see Table 1.2). By contrast, remittances accounted for 35 per cent and non-ODA flows[3] for 48 per cent in in the same year. However, if we look at these flows to the poorest countries (the OECD's definition of 'least developed countries' (LDCs)) the picture is rather different. In 2015, the

Table 1.2 ***Financial flows to developing countries 2000–15***

(a) Total Flows to Developing countries (%)				
	2000	*2005*	*2010*	*2015*
Remittances	23	29	30	35
Non-ODA	59	50	54	48
ODA	18	21	16	17
Total $US mill	449806	643862	899701	1086007
(b) Flows to Least Developed Countries (%)				
Remittances	22	20	29	35
Non-ODA	20	12	15	18
ODA	57	68	57	47
Total $US mill	36018	64311	81270	100208

Notes: gross disbursements on a three year moving average, $US mill at 2015 prices.

Non ODA includes Foreign Direct Investment, other official development flows, export credits, private flows at market prices and private grants.

Data extracted from OECD (2017)

share of ODA was 47 per cent,[4] remittances were 35 per cent and other flows just 18 per cent. In other words, as we should expect, middle income developing countries depend more on market flows and remittances whereas LDCs still rely heavily on aid (and increasingly remittances) for their external financial resource inputs.

Although we see some signs of relative decline in the contribution of ODA to the economies of developing countries as a whole in recent years, it is important to note that it has been a significant feature of the global economy for the past 40 years. Less well-off countries have had to rely on international development assistance as private and market flows of resources from outside have lagged. Also, we can see from Figure 1.1 how ODA has varied over time since 1970. In real terms, adjusting for inflation, ODA broadly grew during the 1970s, plateaued during the 1980s, fell in the 1990s and has grown substantially – basically doubling – since 2000. It has been mainly directed to developing countries in Asia and Africa, though there has been a large growth in the 'unspecified' category in the past decade.

Figure 1.1 ***ODA flows by recipient region 1970–2017***

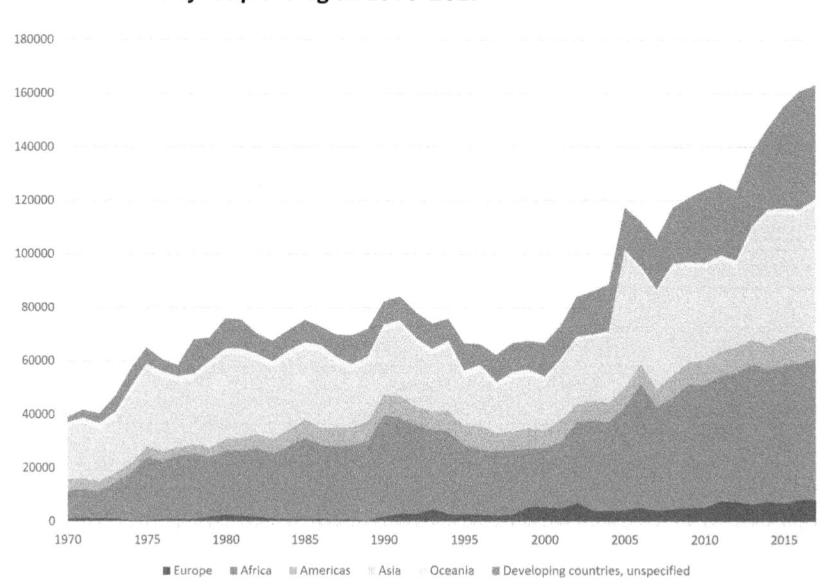

Total official donors, disbursements $US mill constant 2017
Source: www.stats.oecd.org

Over this 50-year period, aid has been part of a series of global 'projects' according to McMichael (2017) – the UN Development Decade launched in 1961, the MDGs of 2000–15 or the present SDGs of 2016–30 – with the stated, and sometimes contested, intentions of promoting development and alleviating poverty on a very large scale. Viewed historically, aid has been the main tool used to transfer resources from richer to poorer countries in the name of these grand projects and ambitious goals.

Justifications for aid

Why should aid be given? Why have donors continued to spend many millions of dollars over several decades on aid projects and programmes often far from their own shores? And why do governments and communities in developing countries continue to receive assistance when there seem to be real doubts about its effectiveness and the conditions that have to be met in return for aid?

Altruism

In some ways, a superficial answer to these questions is very simple: it is altruism – a selfless concern for the well-being of others. Aid does good. It eases human suffering and supports efforts to help people become more prosperous and self-reliant. It is based on a basic human sentiment of generosity – giving makes us feel better individually and collectively – and aims to build goodwill. These are powerful touchstones for aid. They help promote the success of public appeals for donations and allow donor governments to justify the use of taxpayer funds to their electorates. Aid, at its most basic level, does have a high (if variable) degree of popular support. And on the recipient side, aid is gratefully received as it supports efforts by governments and civil society who lack resources to address the pressing needs of their communities. Altruism constructs a dualist model of aid: there are generous and selfless donors on one side; needy and grateful recipients on the other; and there is a one-way flow of resources from the former to the latter.

Yet behind these superficial justifications of aid lie much more complex and contested motives and structures. Altruism may operate at one level to generate support for aid, but effectively it is the self-interest of donors (and recipients) that plays a greater role in shaping the nature and direction of aid and the way it is delivered. There are several different facets to self-interest: economic, political/diplomatic, and welfare, peace and stability have been seen as interacting factors driving Western aid since 1945 (Griffin 1991).

Economic benefit

Economic factors are central to the aid 'industry'. Aid is seen to help promote economic activity and growth – creating employment, providing new goods and services and expanding the economic options available for poorer societies. The provision of capital and technology through aid is seen by many to accelerate investment, growth and employment. This argument is often used to suggest that economic growth ultimately reduces poverty. From the recipient side, aid provides resources that are otherwise in very short supply domestically or too expensive to obtain on the global market. Aid can reduce the need for debt and ease balance of payments problems. For donors, there are also benefits. Economic aid, in the form of loans, capital equipment or technical advice, helps align hoped-for growth in the recipient economy with the trading interests of the donor economy. Investment and trading opportunities open alongside aid, helped no doubt by recipient governments willing to lift restrictions on these for companies from donor countries. And if and when economic growth occurs, consumers in recipient countries are able to buy more imports or seek bigger commercial loans or look to attract more foreign investment. Such uses of aid, then, help construct economic systems and relationships that bring multiplied benefits back to donors in the long run as well as to recipients in the short term. Linked to these economic motives for aid, economic restructuring and trade liberalisation conditions are often attached to aid.

Box 1.2 The Marshall Plan 1948–51

It is tempting to see the origins of present-day aid in the substantial assistance given to Western European countries for the reconstruction of their economies after the Second World War. This aligns with the way the very idea of 'development' was constructed and practiced in the immediate post-war period (Rist 1997).

In 1945, the economies of France, Italy, (West) Germany the UK, Netherlands and Belgium were in poor shape following the destruction and disruption of the war. Keen to help these countries recover and not fall under the influence of the Soviet Union, USA undertook a programme of assistance, the European Recovery

Program (ERP), popularly known as the Marshall Plan (named after George Marshall, the US Secretary of State). In 1947, Marshall declared:

> *It is logical that the United States should do whatever it is able to do to assist in the return of normal economic health to the world, without which there can be no political stability and no assured peace. Our policy is not directed against any country, but against hunger, poverty, desperation and chaos.*
>
> (Reinert and Jomo 2008)

The Marshall Plan was launched in 1948 and for the next four years, USA spent over $US17 billion. Most of this was spent on imports of food, fuel and raw materials from USA and, as the Korean War loomed, much was used to rebuild the military forces of Western Europe. The main recipients were UK, France, West Germany, Italy and the Netherlands. The Soviet Union and its allies were invited to participate but declined. Other features of the Plan included an emphasis on technical assistance, whereby American industrial technologies (and investment) provided models for European industries as they rebuilt. Most aid was in the form of grants (much of these used as credits to buy American imports) but loans were used as well to establish development banking operations.

The Plan was seen as a great success, with West European economic recovery and re-industrialisation given a significant boost (Kunz 1997). Food shortages were all but ended and the Plan seemed to usher in a period of sustained economic growth in Europe through the 1950s. In addition, relative political stability was achieved and the Western alliance was strengthened. Moreover, US exporters benefited greatly as European customers were steered and supported their way.

Although the contexts were quite different – the reconstruction of already established industrial economies compared to the development of largely non-industrial societies – the perception of the success of the Marshall Plan, using American assistance to aid prosperity and strengthen a new American-centred global economy, was to be very important. Similar tools (infrastructure, loans, food aid etc.), a similar large-scale approach and similar goals (modernisation, economic growth and political alignment) could drive a new programme of development assistance to countries in Africa, Asia, Central and South America, and Oceania. Many such countries were becoming independent from their former European colonial masters and were seeking new models for development and new political relationships. Aid could become a key tool in ensuring they 'developed' along capitalist lines and stayed away from the Soviet or Chinese communist blocs. The Marshall Plan provided a template for how aid might work.

Political and diplomatic considerations

There are also political/diplomatic justifications for aid: aid features importantly in the conduct of foreign policy by all sides. Aid donations from one government to another are rarely – if ever –

given without some sort of either overt or tacit understanding that there is an expectation of a political return of some kind. This may involve recipients giving their diplomatic support to a donor's position on the global stage (a vote at an international forum on whaling, a bid for a seat on the United Nations Security Council, or diplomatic recognition or not of Taiwan, for example). In this case, a recipient's status as an independent sovereign state with voting rights is a critical economic resource that can be 'traded' for aid receipts. Aid is also used as an instrument to influence domestic policy in recipient countries. Donors have, in countries such as Kenya or Fiji, withdrawn or threatened to withdraw aid from a government they think lacks democratic legitimacy or has a poor human rights record in the hope and expectation that policies will change. Political and diplomatic gains can be sought by donors in more subtle ways. The giving of tertiary education scholarships to study abroad (in the donor country) has long been a way of building relationships with the future elites in recipient countries and aligning them with donors' values, ways of life and ways of working. Similarly, technical assistance can help reconstruct recipient institutions (customs and immigration, police, drug enforcement, finance etc.) so that they parallel and interact smoothly with those of the donor. Finally aid has a political benefit for donors domestically and internationally. Rising aid budgets help mollify domestic pressure groups calling for action on poverty and debt reduction and internationally, a generous aid budget is a sign that a donor government is a 'good global citizen' which acts responsibly and can be welcomed on the international stage. For emerging economies, making a transition from being an aid recipient to becoming an aid donor (China, India, Chile, Brazil etc.), is also a tangible sign of success and of graduating to join the well-off countries of the world.

Welfare, peace and stability

As well as building economic activity and diplomatic relationships, aid is also given and received to promote welfare development in a strategic and more long-term way. When aid is used to help provide better education and health services, especially when this is done through local institutions, it can bolster public support for governments that donors back, or reduce support for separatist or opposition groups. More peaceful and stable countries with good public services and an expanding economy are also less likely to become homes for subversive and/or violent groups and, as a consequence, sources of refugee flows. In this way, welfare spending offshore by donors is seen as a way of lessening the chance of having to spend more at home (in hosting refugees or combatting terrorism, for example), and is also likely to win popular political support, particularly in polities where populist nationalism now holds sway (such as the USA and UK). Furthermore, more forward-looking donors might see benefits in welfare-related aid projects and the ways this can assist the regulated labour flows. Aid can support particular projects in education and training that will equip future migrants or seasonal workers with the skills to work in donor countries when and if required. In this way, aid can maintain a structure that provides cost-effective social reproduction of a reserve of workers offshore who can then be welcomed and regulated if and when the donor requires.

All these economic, political and strategic justifications for aid thus take us a long way from the simple notion of altruism and the dualistic model of aid relations. Instead of a simple rich-poor divide and a one-way flow of resources, we have a complex world characterised by increasing interdependence through aid and marked return flows of resources and political 'capital'.

Before we leave this initial look at justifications for aid, we should note that not all aid involves government-to-government

Figure 1.2 ***The aid machine***

Source: Cartoon by Polyp reprinted by kind permission.

relationships and this rather cynical view of the hidden motives for aid. Private aid flows, from individuals to and through non-governmental organisations (NGOs) to recipient organisations and communities, and the involvement of wealthy private philanthropists does not accord well with this view of deeper economic and political motivations for aid. A concern for the well-being of others is clearly a key reason people give freely of their own money. These individual altruistic motives flow through to the operations of many development NGOs in both donor and recipient countries. Civil society organisations thus seem to be more attuned to the conditions, needs and aspirations of poorer people and more motivated by concern for them. Yet, even here we should also ask if these are the only motives at play. Some faith-based NGOs are undoubtedly guided by the underlying humanitarian principles of their religion, yet they can also seek to persuade others – recipients – of the virtues of their faith even if just by example. Others may be more explicit, building schools and introducing curricula that are not fully secular, or having proselytising activities alongside aid projects. Other NGOs may use aid not so much as a way of convincing recipients of their values and beliefs but to demonstrate the importance of their broader missions. Many development NGOs have a particular set of objectives – environmental protection, family planning, clean water, domestic violence etc. – and understandably focus their aid projects on these concerns. Overall then, donor governments may use aid as a foreign policy tool and NGOs may use aid as a public advocacy tool. In both cases the worldviews and strategic objectives of donors are as influential, and perhaps more so, than those of the recipients.

Aid critics

There have long been critics of aid and debates concerning its effectiveness and impacts rage through to the present day. While we are not in a position to resolve these debates here, we can outline here some of the main criticisms, from various ideological and theoretical positions, before we turn later in the book to examine them in greater detail.

Early criticisms of aid hark back to the ideas of Thomas Malthus and his followers in the eighteenth and nineteenth centuries with regard to the poverty that arose out of the Industrial Revolution in

Great Britain and Europe more broadly. They believed that assistance for the poor merely led to increased survival rates and higher population growth which used up available resources and therefore perpetuated poverty. This idea was echoed in the 1960s by writers such as the ecologist Garrett Hardin, who used the metaphor of a lifeboat for Planet Earth. This argument suggested that, in the face of disaster, we are not able to help everybody who is adrift in a sea of suffering; by putting more people on the lifeboat (of prosperity) we cause it to sink, thus leading to the demise of all on board (Hardin 1974). In this conceptualisation, tackling poverty actually makes everybody worse off. In a similar line, other less graphic and simplistic views, include those personified by economists such as Peter Bauer. Bauer (1976) argued that aid perpetuated and even magnified poverty, rather than eliminating it. This was because, he believed aid led to governments becoming more powerful, corrupt and inefficient and that individual effort and enterprise would be suppressed. As we will see these ideas eventually played a critical role informing the neoliberal approach to aid and development that is associated with the 'right-wing' of politics (see Chapter 4).

Equally staunch in their criticisms of aid were writers more on the other, 'left-wing', side of ideological spectrum. Terese Hayter in 1971 published a book entitled *Aid as Imperialism* that focused on the work of the International Monetary Fund (IMF), World Bank and United States Agency for International Development (USAID) in Latin America. She argued that donors used aid as a tool to control developing countries in order to further their own economic and strategic objectives. Her work paralleled that of dependency theorists of the time who believed that global trade and capitalism created underdevelopment and poverty in the 'periphery' (often referred to at the time as the 'Third World') as a consequence of the development of wealth in the 'core' (the industrialised economies of Western Europe, North America and Japan). Though poles apart ideologically, Bauer and Hayter both felt that aid did little to tackle global poverty and in fact perpetuated it.

We can see parallels of these views in present aid debates, though the ideological lines are not always quite as stark. There are neoliberal economists who hold to Bauer's views. The Australian Helen Hughes (2003) for example, argued strongly that 'aid has failed the Pacific'. Dambisa Moyo in her book *Dead Aid* (2010) similarly felt that the billions spent on aid in Africa actually worsened the conditions there and led to a dangerous dependence on aid among many governments. Another economist, William Easterly (2006, 2007, 2008; Easterly and Pfutze 2008) has also been highly critical of the way aid has been implemented but he has taken a rather more liberal view, pointing not to macroeconomic distortions but to the way the rights of the poor have been ignored (also Glennie 2008). For him, aid which focuses on meeting the material and technical needs of the poor – mosquito nets or improved sanitation for example – fails to address the fundamental causes of poverty which lie in the lack of individual rights and freedoms of the poor.

Aid supporters

On the other hand, there are vocal proponents of aid. There have been prominent celebrity champions of anti-poverty campaigns, people such as Bono or Bob Geldof (Box 1.3), and these have been associated with campaigns such as Live Aid, Comic Relief, or Make Poverty History. NGOs, such as Oxfam have also been advocates for aid (Oxfam 2010). There have also been wealthy philanthropists such as Bill and Melinda Gates who have put much of their own wealth into aid projects. Perhaps the main academic support for aid, and a critical opponent of Easterly, has been Jeffrey Sachs. His 2005 book, *The End of Poverty,* is an optimistic view of the potential of aid to address the fundamental needs of the poor. To him, more aid, more focus on things such as improved seeds and irrigation, and targeting the combatting of malaria, tuberculosis and AIDS is vital so that people can escape the poverty trap. Aid can tackle poverty and once this is done, better governance and prosperity will follow. Sachs was influential both in measures to meet the MDGs and in the drawing up of the SDGs.

Box 1.3 Concerts, celebrities and aid campaigns

Global celebrities, and in particular music stars, have played a role in the public promotion of aid and poverty reduction campaigns. The notion that musicians should have a political and compassionate voice is perhaps associated with the late 1960s and early 1970s and the rise of the anti-Vietnam war, the civil rights and nascent environmental movements. Acts such as the Beatles, Bob Dylan and Joni Mitchell were often outspoken regarding such issues and John Lennon and later George Harrison became central figures in the peace movement as music and politics increasingly mixed – as it long had done in non-Western contexts such as Latin America and Africa. The Concert for Bangladesh of 1971 was organised by Harrison, encouraged by his friend and master sitar player Ravi Shankar, and included well-known Western musicians such as Eric Clapton, Ringo Starr, and Billy Preston. It was perhaps the first example of a 'global' concert event intended to provide humanitarian relief and promote awareness of development and aid-related issues. The concert sought to provide relief to East Pakistan, now Bangladesh, following the independence war of 1971 and the related genocide coupled with environmental catastrophe following a cyclone and floods. This concert set the tone and in the following decades such events grew in frequency and magnitude. Perhaps the most iconic of all these events was Live Aid of 1985 organised principally by Bob Geldof, a singer with the well-known post-punk band from Ireland, the Boomtown Rats. At the time this was the largest ever such event involving a global satellite broadcast from both sides of the Atlantic drawing in dozens of the most popular international musicians of the time to raise money for famine relief in Ethiopia. The event itself followed the recording and release of the Band Aid charity single 'Do They Know Its Christmas Time?' organised by Geldof and Midge Ure (of Ultravox) which became the highest selling single of all time in the UK at the time and set a model for charity recordings across the world (such as 'We Are the World' in the USA) over subsequent decades.

Bob Geldof and Bono of U2 became increasingly associated with political campaigns in the wake of Live Aid, culminating in the co-foundation of the ONE foundation to end extreme poverty in the case of the latter. The Live Aid model was repeated in 2005 as Live 8 (named after the then forthcoming G8 summit) and was the flagship event of the Make Poverty History campaign. At this point several other figures, including Chris Martin of Coldplay and a number of screen actors, became involved in what had grown into a broader global justice movement arising in part out of the 'anti-globalisation' movement of the late 1990s / early 2000s. This movement found a political counterpart in the UK Labour Party. Chancellor Gordon Brown had lobbied to cancel the debts of the most highly indebted countries in the Global South based in part on the work of the social movement / NGO Jubilee 2000 which had had some success in convincing the IMF and World Bank to undertake such reform. Based in part on this groundswell in political concern and public approval aid per capita in the UK and some other OECD countries rose in the mid-2000s until the middle of the 2010s. The Make Poverty History continues to lobby governments and raise public awareness at the time of writing and played an instrumental role in the formation of the United Nations SDGs of 2015 seeing the goal of 'no poverty' by 2030 being placed at number one.

There has been criticism of the role of celebrities, which can be perceived as self-serving (see Brockington 2014; McVeigh 2017). The absence of artists from Africa and Latin America at global events such as Live Aid and Live 8 has sometimes been pointed to as evidence of a condescending, and at worst neo-colonial, approach to solving development problems. Furthermore, events – especially in the earlier days – were approached from a depoliticised point of view blaming famine on environmental processes rather than political factors. This contributed to a popular misconception in the West, promoted by media, and complicit governments, that underdevelopment was caused by natural events and overpopulation. As such, celebrity involvement often cited 'emergency' responses to social and environmental catastrophes rather than highlighting the underlying structural causes. It seldom, if ever, cited the development policies of the West as instrumental in creating underdevelopment and associated famine in the first place. This changed with the rise of the Make Poverty History Campaign but the involvement of governments and mainstream celebrities often led to accusations of the watering down of the radical message of the movement. Notwithstanding such criticisms, relief concerts have raised public awareness greatly and precipitated the rise of larger-scale anti-poverty movements as well as perpetuating academic and policy concern for aid and other development interventions. Celebrities have most certainly helped raise awareness of the issues involved.

Thus, we are faced with a world that does not agree on the motivations, rationale, impacts and even need for aid. Academic and popular debate continues and there are growing schisms in terms of donor policy. While aid has generally increased steadily in the past 20 years (Figure 1.1), there are signs in recent political changes in USA and UK in particular, that governments want to reduce aid and focus more on their own economies. We seem to be at a critical juncture with aid: will we subscribe to the optimism of Sachs, pursue the new targets of the SDGs and continue to sustain aid increases; will we heed the criticisms of Easterly and Moyo and seek to reduce or radically transform aid; or will we follow the increasingly inward-looking foreign policy signals of populist nationalism such as that which currently exists in USA and reduce and re-direct aid budgets?

Box 1.4 Different political perspectives on aid

Two short extracts made by North American leaders tell us a great deal about the varying views about the role of aid as well as how this fits with their view on how they relate to the rest of the world. There is no consensus on such matters and viewpoints alter with political cycles. While there may appear to be an increasing awareness and acceptance of the obligations of wealthier countries to embrace the relationships that are embodied in the SDGs this is by no means guaranteed into the future.

Donald Trump's State of the Union Address 30 January 2018

The United States is a compassionate nation. We are proud that we do more than any other country to help the needy, the struggling, and the underprivileged all over the world. But as President of the United States, my highest loyalty, my greatest compassion, and my constant concern is for America's children, America's struggling workers, and America's forgotten communities. I want our youth to grow up to achieve great things. I want our poor to have their chance to rise. ...

Last month, I also took an action endorsed unanimously by the Senate just months before: I recognized Jerusalem as the capital of Israel.

Shortly afterwards, dozens of countries voted in the United Nations General Assembly against America's sovereign right to make this recognition. American taxpayers generously send those same countries billions of dollars in aid every year.

That is why, tonight, I am asking the Congress to pass legislation to help ensure American foreign-assistance dollars always serve American interests, and only go to America's friends.

Trump (2018)

Justin Trudeau Address to UN General Assembly 21 September 2017

The Sustainable Development Goals are as meaningful in Canada as they are everywhere else in the world, and we are committed to implementing them at home while we also work with our international partners to achieve them around the world.

This is important, because poverty and hunger know no borders. We cannot pretend that these solvable challenges happen only on distant shores. ...

Internationally, we have reaffirmed Canada's commitment to reducing poverty and inequality, putting gender equality and the empowerment of women and girls at the heart of our development efforts.

We took this approach because we know that when we empower women and girls, economic growth follows.

Peace and cooperation takes root.

And a better quality of life for families and communities is possible.

Trudeau (2018)

Conclusion

This brief introduction to aid and development has opened the door to a myriad of questions and issues to explore. What might have started as a simple dualistic model has been revealed to be very complex and contested. We know that aid has resulted in very large commitments of funds over several decades and this has been done in the name of humanitarian relief, helping others and promoting development. But we can see that the reasons aid is given run much deeper than this, there is no simple linear flow of resources from rich to poor, and there is not even agreement on whether aid actually does any good!

Summary

In this chapter we:

- Outlined the contents of this book and the major questions that we seek to explore. These included questions on the definition of aid, its geography, and history as well as how it is delivered, whether it works and what the future may hold.
- Discussed what is meant by aid and how definitions and therefore measures are often critically challenged and yet crucial to understand in given contexts.
- Discussed the link between the current SDGs and how these relate to current and historic aid practices.
- Explored briefly motivations for aid. We saw that justifications frequently included both altruistic and strategic motives.
- Introduced the idea that aid can be given for economic, political/ diplomatic, and security reasons in terms of strategic motives and noted that these factors overlap in different ways in various places and over time.
- Outlined the broad contours of the debate concerning critics of aid from both the right and left wings of politics.
- Introduced outline arguments that support the use of aid from both a political and academic point of view.
- Looked at the role of celebrities and public campaigns in the public perception of what aid is and what it is for.

Discussion questions

- What are some of the main justifications for giving aid and how can these overlap?
- How do we define aid and how is it measured?
- Discuss the main arguments for and against aid.
- How do the 2015 SDGs relate to the concept and practice of aid?

Websites

- Does aid work?: https://www.oxfam.org/en/multimedia/video/2010-does-aid-work
- The SDGs: https://sustainabledevelopment.un.org/sdgs and https://sdg-tracker.org/

Videos

- Ed Sheeran: https://www.youtube.com/watch?v=LHKDAF9XKoo
- Ricky Gervais: https://www.youtube.com/watch?v=5DgIRjecItw
- Africa for Norway: https://www.youtube.com/watch?v=oJLqyuxm96k

Notes

1 In this book we sometimes use the terms 'developing' and 'developed' to refer quickly to countries with varying levels of income per capita and living standards. We are aware that at best these are imperfect terms that hide more than they reveal and at worst they actively discriminate and reproduce inequality. We also use the terms Global North and Global South. In addition, we draw on the further classification of 'developing' countries used by the World Bank and OECD. This has sub-categories of 'low income' (or 'least developed'), 'lower-middle income' and 'upper middle income' countries, as well as 'high-income' (developed) countries (see for example OECD 2019a).

2 Measured here as ODA (Official Development Assistance) – see Chapter 2.

3 Non-ODA flows are comprised of 'Other official development flows, officially-supported export credits, FDI, other private flows at market terms and private grants' (OECD 2017).

4 ODA had been as high as 68 per cent in 2005.

Further reading

Collier, P. (2007) *The Bottom Billion: Why the Poorest Countries Are Failing and What Can Be Done About It.* Oxford University Press, London and New York.

Easterly, W. (ed.) (2008) *Reinventing Foreign Aid.* MIT Press, London and Cambridge.

Glennie, J. (2008) *The Trouble with Aid: Why Less Could Mean More for Africa.* Zed Books, London.

Mawdsley, E., Savage, L. and Kim, S.M., 2014. 'A "post-aid world"? Paradigm shift in foreign aid and development cooperation at the 2011 Busan High Level Forum,' *The Geographical Journal*, *180*(1), 27–38.

Moyo, D. (2010) *Dead Aid: Why Aid Is Not Working and How There Is Another Way for Africa.* Farrar, Straus and Giroux, New York.

Oxfam (2010) *21st Century Aid: Recognising Success and Tackling Failure.* Oxfam Briefing Paper 137.

Sachs, J.D. (2005) *The End of Poverty: Economic Possibilities of Our Time*. Penguin, New York.

2 What is aid?

Learning objectives

This chapter will help readers to:

- Understand that the concept of aid is broad and contested. It is more complex than definitions utilised by the Organisation for Economic Cooperation and Development (OECD)
- Appreciate that aid worldviews have shifted over time, although all to an extent continue to exist and overlap in different places at different points in history
- Assess how views concerning aid have shifted from simplistic notions of altruism and giving to complex multi-faceted, strategic and sometimes mutual interest-based sets of policies
- Understand that official definitions of aid have changed over time to include a wider range of forms and yet still does not include everything that might be reasonably considered 'aid'
- Assess how the geography of aid donors and recipients has shifted markedly over time as the line between 'developed' and 'developing' shifts and the line between donor and recipient becomes blurred
- Understand that there are many forms of aid beyond traditional Official Development Assistance (ODA) as measured by the OECD
- Evaluate the rise of new aid powers, such as China, and understand the diverse strategies and forms their assistance is taking
- Discuss the concept of 'South–South' co-operation and the notion of a post-aid world

Aid is one of the most prominent features of debates concerning both development and international relations. Flows of aid account for many billions of dollars across the globe each year, they tie countries together in webs of negotiations and agreements, and they contribute to the improvement in the welfare of many millions of people. There seems to be a strong moral justification for giving aid (Culp 2016). Yet despite the significant size of aid flows and continuing efforts to promote development worldwide, we are still rather in the dark regarding what is meant by 'aid', how we might define and measure it, let alone understanding whether or not it meets its objectives.

The word 'aid' implies a number of things. As we saw in Chapter 1, it invokes notions of 'help' and 'assistance'. In doing so, it suggests that aid relationships involve a net flow of resources from one party (a donor) to another (a recipient). It also implies a sense of altruism: that assistance is given for the sake of 'doing good', of helping those in need, and for donating resources without the need for an equivalent return. 'Aid' is thus a powerful discursive tool for it suggests to us that the world is structured in particular ways with certain inequalities in power and resources, and flows which aim to meet some of the most pressing of humanity's needs. As discussed in Chapter 1, in this book we will explore the complexities of international aid and see how it is linked to different ideas of 'development'. In this chapter, we examine the way the term aid is defined and measured. Here we steer the discussion towards what has been the dominant architecture of aid, that centred on the OECD/DAC definition of ODA and constructed by a dualistic definition of donors and recipient countries. We deconstruct this mainstream view of aid by firstly describing forms of assistance that have not been included in ODA (such as migration or trade preferences) and then identifying new players and evolving 'South–South' ways of operating that are beginning to challenge and disrupt the OECD-centred aid hegemony.

Changing aid worldviews: motivations through history

Aid has changed over time to become a very diverse and complex set of activities and resource flows. In what follows we consider four 'aid worldviews' in the evolving rationale and nature of aid: 1) helping those in immediate need; 2) long-term modernisation and welfare improvement; 3) advocacy and emancipation through education; and 4) shared prosperity and explicit mutual benefits. Although elements of all of these sets of motivations persist and overlap, they have, in rough terms, superseded one another chronologically.

(i) Aid as relief

Firstly, at its most basic, aid has involved the provision of financial and material resources to those 'in need'. If we look at the origins of many development agencies, be it the World Bank or Oxfam, we see this most basic function as a key starting point. The World Bank, for example, had an initial mission to focus on reconstruction from the chaos in Europe following the Second World War. Oxfam started as a famine relief organisation focused on Greece in 1942. Disasters and humanitarian relief provide the most obvious *raison d'être* for assistance: people suffer because they lack food, shelter, clothing, potable water and the like and their survival is threatened unless external resources can be provided. This justification for aid continues very much through to the present, particularly in the public imagination of aid. Natural disasters – droughts, tsunamis, floods, earthquakes etc. – are followed by calls for donations and the sending of materials to affected areas. Disasters can also be human-induced – people are displaced by war and conflict, insecurity, forced resettlement, economic collapse – and similar responses may be forthcoming.

The concept of aid in this sense appears to be simple. People are in dire need with their welfare and very survival threatened and other communities have surplus resources which they can donate to alleviate human suffering. People in relatively well-off countries and communities often respond readily to images of starving children or destitute mothers or devastated homes. The assembling and dissemination of these images accompanied with appeals to give has constructed a particular public imaging of aid: it is one-way, it is based on pressing need, and it is effective in relieving suffering.

(ii) Aid as development

Yet whilst this basic view of aid discussed above persists, it has been joined – and superseded in influence – by a second set of motivations, which sees aid as a way of underwriting forms of development that pre-empt suffering and enhance human welfare. The development rationale for aid is long-standing and can be traced most explicitly back to the 1940s and 1950s and ideas of modernisation. Some (e.g. Rist 1997; Escobar 1995) trace it directly to President Truman's inaugural address in 1949 (Box 2.1).

Box 2.1 President Truman and the post-war Western justification for aid

In his inauguration address in 1949, President Harry Truman established a justification and agenda for development assistance in the post-Second World War and impending Cold War international environment:

> *... we must embark on a bold new program for making the benefits of our scientific advances and industrial progress available for the improvement and growth of underdeveloped areas.*
>
> *More than half the people of the world are living in conditions approaching misery. Their food is inadequate. They are victims of disease. Their economic life is primitive and stagnant. Their poverty is a handicap and a threat both to them and to more prosperous areas.*
>
> *For the first time in history, humanity possesses the knowledge and the skill to relieve the suffering of these people.*
>
> *The United States is pre-eminent among nations in the development of industrial and scientific techniques. The material resources which we can afford to use for the assistance of other peoples are limited. But our imponderable resources in technical knowledge are constantly growing and are inexhaustible.*
>
> *I believe that we should make available to peace-loving peoples the benefits of our store of technical knowledge in order to help them realize their aspirations for a better life. And, in cooperation with other nations, we should foster capital investment in areas needing development.*
>
> *Our aim should be to help the free peoples of the world, through their own efforts, to produce more food, more clothing, more materials for housing, and more mechanical power to lighten their burdens.*
>
> *We invite other countries to pool their technological resources in this undertaking. Their contributions will be warmly welcomed. This should be a cooperative enterprise in which all nations work together through the United Nations and its specialized agencies wherever practicable. It must be a worldwide effort for the achievement of peace, plenty, and freedom. (Truman 1949)*

Truman's vision then was for the direct use of the Western world's technology and materials, through aid and trade, to promote economic development and poverty alleviation in poorer parts of the world. Recipients – 'more than half the people of the world' – would be linked in a new world economic order led not by the old European-controlled colonialism, or evolving communism, but by the industrial might of USA. It was not the direct manifestation of severe poverty and suffering (as with disaster relief) that led to aid but rather the construction of it as an idea that was associated with 'misery'. The use of adjectives in Truman's speech, and other documents of the time, such as 'inadequate', primitive' and 'stagnant' depicted non-Western lifestyles and livelihoods as in need of intervention and assistance. By contrast the modern, progressive, wealthy and democratic nations of the West would provide the templates for aid-supported recipients to follow and copy. Note the use of the term 'free peoples' and 'peace loving'. This was a direct rejection of assistance to communist and communist-aligned states such as those in Eastern Europe, and later parts of Asia, Africa and Latin America. In this sense the dawn of 'development', and the aid it gave rise to, had an explicit geopolitical objective to prevent the spread of the 'evil-empire' of communism. In this sense it was a central part of the evolving Cold War principally between the USA and the Soviet Union and allies on both sides where the proxy war was often played out.

We can see clear parallels here with the thinking behind the Marshall Plan (Box 1.2) that was operating at the time and had a similar view of the USA's superior technology and economic performance and a similar goal to reconstruct a new post-war global political and economic system centred on, and dominated by that country. Truman laid this foundation for the grand post-war 'development project' (McMichael 2017), which had aid at its centre: aid that would provide not just material resources but perhaps more importantly the 'technical knowledge' to align their development aspirations and trajectories with those of the West. This development rationale for aid was adopted through the 1950s and 1960s by major aid donors (USA, UK, France, Japan etc.), by the large multilateral agencies (the World Bank, the United Nations) and later by NGOs who shifted from humanitarian relief to development activities to improve long-term human welfare.

Aid for development involved the full gamut of what we consider development assistance: loans for hydroelectric dams or roads; the building of schools and hospitals; improving rural water supplies and irrigation; technical assistance with better crops; building the capacity of local officials to carry out the modern functions of a state; scholarships to study abroad; better urban sanitation; provision of medicines; public health and family planning campaigns; and a plethora of other projects and programmes. In terms of public support, this system of aid was not as simple to justify as it meant long-term commitments to needs that were not quite so obvious or pressing as a famine or an earthquake, for example. Simple and facile arguments such as 'give a poor man [sic] a fish and you feed him for a day; you teach him to fish and you give him an occupation that will feed him for a lifetime' could be used to justify such spending. Yet the approach took donations less directly out of the pockets of members of the public and more from the public accounts of new government aid agencies: aid became a function of the state rather than being left to the vagaries of public choice. This move linked aid to the wider strategic and diplomatic considerations of donor states and there, to a large extent, ODA has resided ever since.

(iii) Aid as advocacy

A third less-direct, non-material rationale for aid has evolved over time but its origins and impacts are less obvious. Advocacy is a much less tangible form of assistance, yet its proponents would argue that it is just as important, if not more so, than providing material resources. Advocacy stems from the ideas of Paulo Freire (1970) and others who saw that poverty was linked to forms of oppression and this was best tackled through people becoming aware of the systems which impoverished them so they would be empowered to take action themselves. Aid in this emancipatory sense becomes a means to support education and public awareness and it can have a quite radical and destabilising effect as it attempts upset the status quo of power imbalances, subjugation and impoverishment. We have seen this advocacy approach used in relation to gender-based aid and development interventions. Here aid is used to support both public awareness programmes (for example, against gender-based violence) and particular programmes that offer training and support and engage in political lobbying on women's issues.

Awareness raising and advocacy for the rights and actions of the poor also involves working with people in donor as well as recipient countries and, often, promoting ways of linking them together. This can have a political dimension, encouraging donor governments to recognise and support the efforts of, say, a particular minority group in a partner recipient country. Or it can be economic, using fair trade networks to link producers and consumers. Because of its political and transformative agendas, aid for advocacy often faces suspicion on the part of donor agencies (though historically some European donors have been active in this manner) and sometimes active resistance on the part of recipient states. Advocacy, though, does require aid activities to consider the underlying causes of poverty and support action to tackle those causes, rather than accept that poverty alleviation can be addressed simply by giving more resources or setting countries on a particular path of trade, economic growth and modernisation.

(iv) Aid as donor self-interest

We might add another set of motivations and resultant view of aid that has evolved in recent years, that which seeks to use aid explicitly as a way to promote the interests of donor economies (alongside those of recipients). It is important to note that aid donors perhaps have always sought some form of return, economically, diplomatically or in terms of their international reputation. Yet in the past decade especially, we have seen donors become a lot more open about what economic benefits they expect from an aid programme. This has been encapsulated in the 'shared prosperity' mantra (more on that in Chapter 3) and seen in the way the People's Republic of China (PRC) has preferred to talk about South–South co-operation rather than 'aid'. There are now more overt efforts to use 'tied' aid to link the business operations of donor companies with development activities offshore. This also aligns with efforts to broaden the government base of aid delivery and calls for a 'whole-of-government' approach to ODA, counting the offshore operations of different line ministries (police, defence, customs etc.) within aid budgets.

(v) Aid as private sector development

As part of this recent re-conceptualising of aid, we have seen an explicit effort to widen the arena of aid to see not just government agencies or NGOs as the key players and improved welfare as the key objective, but also to include the private sector as a crucial driver of development and economic growth, investment and employment creation as aspects which can be supported by aid. The Busan High Level Forum on aid effectiveness in 2011 was a notable event where these ideas were expressed and this was backed-up by the Addis Ababa meeting on financing for development in 2015, which argued that private sources of capital should be included in the way development and aid are conceived and practiced (United Nations 2015a).

This shift to include the private sector has a number of important implications for aid. Firstly, we can see how new terminology and activities have entered the realm of aid. Thus, we have seen how private philanthropists (such as the Gates, Bloomberg, Zuckerberg, Bezos and Buffet family names) have become much more part of the public consciousness of aid, largely because the sheer volume of their philanthropy (Bill and Melinda Gates are reported have donated over $US35 billion – Kumar 2019: 21) now ranks them ahead of many donor countries in the ODA rankings. We also see talk of 'social enterprises', 'corporate social responsibility', crowdfunding' and 'brand aid' as ways whereby new sources of finance are tapped through market enterprises to be made available for development and welfare projects (Kumar 2019; Richey and Ponte 2011). More fundamentally, we are seeing changes in the way we conceive forms of development. As we will see in Chapter 4 and the discussion of 'retroliberalism', there has been a marked discursive shift in the past decade in the way development is conceived at the policy level. Economic growth – rather than the prior Millennium Development Goal (MDG)-inspired focus on welfare (poverty alleviation, education, health, gender equity, etc.) – is seen as the most effective way to achieve development and it is assumed that if there is growth and the creation of new enterprises and jobs, then development has been achieved. This then has reconstructed the view of aid. Aid, it is believed, can promote economic growth by supporting the private sector to invest and the market, assisted by donor aid strategies, will ensure that increasing employment and incomes will alleviate poverty and ensure improved well-being for all. Unlike the neoliberalism of the 1980s and 1990s, when the market was left alone to operate and aid was reduced, this new aid environment openly lobbies for aid to be increasingly channelled to support and promote private sector development.

Aid motivations: summary

So, we have seen that aid has progressed from a simple notion of giving to those in dire need to incorporate strategies for long-term development, support for self-determination and recently for a sort of shared programme that facilitates economic growth on both sides of the aid relationship. These 'aid worldviews' as we term them have then become formalised and evolved sets of guiding principles and methods of delivery that can be termed 'regimes', as we later discuss in more detail in Chapter 4. We can question whether the general change we have witnessed can be considered 'progress' in aid at all. Given the most recent worldview that has turned to concepts of shared prosperity some have suggested the concept of aid is no longer relevant, we have moved so far from the simple notion of 'aid' to a point where perhaps the term has lost its utility. It is now so muddled with competing motivations and strategies and so diluted that perhaps we are approaching what Mawdsley *et al.* (2014) have termed a 'post-aid world'.

Defining aid: Official Development Assistance (ODA)

Given these contested and dynamic conceptualisations of aid, we can now turn to the way 'aid' is defined. The issue of defining aid has attracted the attention of donors in particular. Donor countries, particularly through the agency of the Development Assistance Committee (DAC) of the OECD have attempted to standardise the way aid is defined and how it can be measured (Box 2.2). The DAC has acted as the main institution collecting data on aid flows and share common practices. The DAC approach to defining aid has been to employ the term Official Development Assistance (ODA) in order to more precisely frame what aid is or should be. ODA is an important term as its three components mark significant aspects of aid: 'official' means that ODA only flows through recognised (usually state or large multilateral) agencies; 'development' means that it must be used for funding activities that promote forms of development-related activity (rather than, say, military expenditure); and 'assistance' implies that there should be a net flow of benefits from donor to recipient.

The DAC definition of ODA therefore is:

> *'those flows to countries and territories on the DAC List of ODA Recipients and to multilateral development institutions that are:*
>
> i. provided by official agencies, including state and local governments, or by their executive agencies; and
> ii. each transaction of which:
>
>   a) is administered with the promotion of the economic development and welfare of developing countries as its main objective; and
>   b) is concessional in character'.[1]
>
> (OECD 2019b)

Box 2.2 Key institutions: the Development Assistance Committee of the Organisation for Economic Cooperation and Development (DAC/OECD)

The DAC of the OECD and its secretariat the Development Co-operation Directorate (DCD-DAC) perform a very important series of functions within the OECD system. It has been critical over time in developing and disseminating amongst its members changing approaches to aid (Eyben 2013). An initial Development Assistance Group (DAG) was established in 1960 and this was formalised with the establishment of the DAC the following year with 11 founding members. DAC resolved that it should 'Agree to recommend to members that they should make it their common objective to secure an expansion of the aggregate volume of resources made available to the less-developed countries and to improve their effectiveness' (DAC 2006: 10).

From the outset, the DAC embarked on a process of peer reviews with committees drawn from member countries examining the aid programmes of other members and making recommendations accordingly. These peer reviews continue today and are effective means of sharing and reviewing practices within the donor community.

The Committee was also keen to lobby for increases in aid budgets and as early as 1964 set the target of devoting 1 per cent of national income to development assistance (this was later changed to 0.7 per cent). In this sense, DAC perhaps can be seen as an advocate for development assistance, its more progressive members particularly in Northern Europe pressing through the Committee for other OECD members to become both more generous in their aid budgets and more consistent in the way aid principles were applied.

In 1969, it adopted the ODA measure as a more precise way of measuring aid flows. The DAC's role in collecting data has been a major function ever since. It has required members to supply statistical data on aid programmes, it has applied standards to what can be included or not (though these have been open to debate and change), and it has analysed and published these data, helping to make transparent the large and complicated financial and other flows from members to recipient states.

Another key function of DAC has been the way it has promoted what has been regarded as good practice. For example, in the 1990s it helped encourage donors to adopt specific programmes to promote the concerns and needs of women in development and it also took a lead in asking members to consider the environmental impacts of development activities at the time. Although neither of these was at the cutting edge of critical thinking on these matters, and some bilateral donors had more progressive approaches, DAC did play an important role in encouraging some member donors (through its peer reviews and policy papers) to think about and adopt such elements in their aid strategies.

Perhaps the major achievement of DAC in this regard has been its promotion of the aid effectiveness agenda. DAC initiated the international meetings on aid effectiveness starting with the Rome first high-level forum in 2003. The subsequent

Paris Declaration of 2005 became a cornerstone of aid policies and practices thereafter and did much to steer aid away from tied aid and donor-led approaches (see Chapter 3).

Membership of the DAC will continue to expand (Table 2.2) and although it often takes on a more progressive stance with regard to aid principles, it remains a reflection of the interests and policies of its constituent members.

Although this seems a precise definition, in practice, DAC has modified what can be counted or not over time and there are contentious areas on the margins of this definition. For example, although there is a strong statement that military expenditure is excluded from ODA, the costs incurred by the armed forces of donors to deliver humanitarian assistance may be counted as can some 'developmentally relevant' peacekeeping activities by armed forces. Nuclear energy for civilian uses can be counted also (OECD n.d.). In one of the most bizarre episodes in the changing definition of ODA, there has been debate concerning what forms of military expenditure might be included with regarding to peacekeeping (DAC Secretariat 2016a). Donors such as the United Kingdom have preferred a wider definition of ODA to include 'peace and security costs and well as the costs of countering violent extremism' (Devex 2016) whilst others, such as Sweden have resisted this potential militarisation of aid. What has happened is that DAC has resisted some forms of military aid accounting such as military equipment or salaries but some forms, such as 'expendables' (perhaps even ammunition!) and training, may be allowed (Heinrich *et al.* 2017; Christie 2012).

Recently, new elements have been added to the list of what counts as ODA. For example, the costs of refugee resettlement within a donor country in the first year are now included and this has led to a significant widening of the scope of ODA accounting (and subsequent favourable impression) for those countries receiving and supporting large numbers of refugees. These costs have become a substantial element of ODA reporting: 'In 2015 there were 10 members [of OECD-DAC] for whom in-donor refugee costs were between 10% and 34% of total ODA' (OECD 2016) and this was particularly the case for European donors. There are pressures to widen even further the ODA definition, to include, for example, donor government support for donor country companies to do business in developing countries. There are complex issues regarding more diverse financial instruments and the degree these are concessionary or not, and there is a thorny question of 'additionality' ('additional' funds spent by the private sector in the course of normal business, but which address the development and welfare of a country). To a large extent, these changes have not only made aid accounting much more complex, but also taken the realm of ODA away from dedicated aid agencies within donor governments. Rather, a wide range of activities is now accounted for involving activity at 'home' and abroad by many donor state agencies including the military, police, many government departments, and private sector.

ODA, then, is a useful measure for us in order to glean an idea of what forms of assistance flow from some countries to others. Furthermore, the DAC framework gives us a standardised, if imperfect, approach to defining and measuring aid. It is far from ideal as its criteria have changed over time and it does not include a number of significant forms of assistance (see later in this chapter) and some significant donors (such as the PRC). Also, it blinds us to significant reverse flows from 'recipients' to 'donors' (such as political favours, economic concessions, loan repayments, profits and investment dividends etc.) and the way these may be linked to aid receipts. However, its core elements (official, development and assistance) specify important principles and its approach to data collection gives us a reasonably consistent, if constrained, set of data to track and analyse aid flows.

Defining aid: recipients and donors

Defining and tracking ODA over time is also complicated by changes in counting what might be considered a 'developing' country (and be regarded as a recipient) and who is seen as a donor. The DAC largely draws on the World Bank's classification of low-, middle- and high-income countries. The list of recipients includes all those classified as least-developed countries and all low- and middle-income countries (except any who are members of the G8 or European Union). Those that exceed the high-income level for three years in a row are removed. Looking at changes in countries classified as recipients since 1989 (Table 2.1) we can see some major shifts. Of the 24 countries added to the list, most are the result of the dissolution of the former Soviet Union and Yugoslavia into new states (as well as the splitting of the former US Trust Territories of the Pacific Islands into four Micronesian states). However, those 42 countries that have fallen off the list represent some examples of apparently successful economic

Table 2.1 ***ODA recipients: changes over time since 1989***

Countries added:	*Countries removed:*
Albania (1989) South Africa (1991)	Portugal (1991) French Guyana (1992)
Kazakhstan (1992)	Guadeloupe (1992)
Kyrgyzstan (1992)	Martinique (1992)
Tajikistan (1992)	Réunion (1992)
Turkmenistan (1992)	Saint Pierre and Miquelon (1992)
Uzbekistan (1992)	Greece (1995)
Marshall Islands (1992)	Bahamas (1996)
Federated States of Micronesia (1992)	Brunei (1996)
Armenia (1993)	Kuwait (1996)
Georgia (1993)	Qatar (1996)
Azerbaijan (1993)	Singapore (1996)
Eritrea (1993)	United Arab Emirates (1996)
Bosnia and Herzegovina (1993)	Bermuda (1997)
Croatia (1993)	Cayman Islands (1997)
Macedonia (1993)	Chinese Taipei (1997)
Yugoslavia – Serbia and Montenegro (1993)*	Cyprus (1997)
Slovenia (1993)	Falkland Islands (Malvinas) (1997)
Palau (1994)	Hong Kong (China) (1997)
Northern Marianas Islands (1994)	Israel (1997)
West Bank and Gaza Strip (1994)	Aruba (2000)
Moldova (1997)	British Virgin Islands (2000)
Kosovo (2009)	French Polynesia (2000)
South Sudan (2011)	Gibraltar (2000)
	Korea (2000)
	Libya (2000)
	Macau (China) (2000)
	Netherlands Antilles (2000)
	New Caledonia (2000)
	Northern Marianas Islands (2000)
	Malta (2003)
	Slovenia (2003)
	Bahrain (2005)
	Saudi Arabia (2008)
	Turks and Caicos Islands (2008)

Countries added:	*Countries removed:*
	Barbados (2011)
	Croatia (2011)
	Mayotte (2011)
	Oman (2011)
	Trinidad and Tobago (2011)
	Anguilla (2014)
	Saint Kitts and Nevis (2014)

Note: * Serbia and Montenegro were listed separately in 2006.

growth: the oil-rich states of Brunei, Bahrain, Kuwait, Saudi Arabia, UAE and Qatar; the Asian 'tiger' economies of Singapore, Korea, the Republic of China (Chinese Taipei) and Hong Kong; a multitude of small island states in the Pacific and Caribbean that have benefited from tourism, remittances and the like; and several European countries that have joined the European Union, such as Portugal and Greece.

However, the DAC recipient list does not give a fully accurate picture of who receives aid. Leaving aside the question of internal forms of assistance (spending by governments on poorer regions or groups within their own countries), we know that some donor governments continue to provide support for other countries even after they graduate from middle-income or recipient status. In the South Pacific, for example, French ODA to French Polynesia and New Caledonia amounted to two of the three largest flows of aid to the region prior to 2000 (alongside Australian aid to Papua New Guinea). Yet in that year, the two French territories were removed from the list of recipients mostly because estimates of their income put them in the high-income category, but perhaps also because there was a political statement to be made that they were the responsibility of metropolitan France and an 'internal' matter for funding, rather than fully independent states receiving international development assistance. We know that aid did not stop to these territories after 2000 and very large flows continue in the form of budget support and development projects (Prinsen *et al.* 2017; Overton *et al.* 2019). However, these are not recorded as ODA by DAC.

On the donor side, there have also been significant shifts. Membership of the OECD and the DAC (not all members of the OECD join the DAC) has expanded. Table 2.2 shows that the 11 original members of the DAC in 1961 were joined by another 10 by 2000 and 8 since (the latter reflecting changes in the political geography of Europe). Some

Table 2.2 *DAC members*

Original Members (1961)	*Later Members (year joined)*
Australia	Norway (1962)
Belgium	Denmark (1963)
Canada	Sweden (1965)
European Union (& predecessors)	Austria (1966)
France	Switzerland (1968)
Germany	New Zealand (1973)
Italy	Finland (1975)
Japan	Ireland (1985)
Netherlands	Spain (1991)
UK	Greece (1999)
USA	Hungary (2004)
	Korea (2010)
	Czech Republic (2013)
	Iceland (2013)
	Poland (2013)
	Portugal (2013)
	Slovak Republic (2013)
	Slovenia (2013)

countries that had originally been classified as recipients (Portugal, Greece, Korea) have gone on to join the DAC and become donors themselves and, as we shall see, there are other countries that remain on the list of recipients and continue to receive ODA (for example, India and Indonesia), but who have become donors themselves.

As the global economy has changed, its old North–South dual structure has weakened considerably and, with that, we have seen a blurring of the aid donor-recipient distinction. Not only are countries graduating from 'developing' to 'developed' status, they are also shifting from being recipients to being donors. However, in the process of transition from one to the other, we are seeing interesting new aid relationships and practices evolving. In the case of Latin America, for example, it is evident that some countries, such as Brazil and Chile (Box 2.3), are developing their own aid strategies, even whilst they may still be receiving some forms of ODA. Other countries in this category who have sought membership of the OECD, but who are not yet members of the DAC include Estonia, Israel and Slovenia (newer members of the OECD) and Russia, Lithuania, Colombia and Costa Rica (seeking accession to the OECD).

Box 2.3 Emerging aid donors: Chile

The example of Chile is an interesting illustration of how countries may undergo a transition from recipient to donor, or hold both positions simultaneously. Chile has always been included in the DAC list of recipients and has received aid in ebbs and flows over the past 50 years (Figure 2.1). High inflows of aid in the 1960s were followed by a decline after Pinochet's military coup in 1973. Although ODA fell, the regime continued to receive military aid, notably from USA and ODA increased again in the late 1980s despite continued human rights abuses. The end of military rule in 1990 ushered in further increases, but thereafter it has fluctuated. Despite a rapidly growing economy over the past 30 years, based on mineral exploitation, Chile has continued to receive about $US 100 mill in ODA per year (in today's currency), though this has fluctuated from year to year. Its main donor, recently, has been Germany.

Figure 2.1 ***Chile ODA receipts 1965–2017 (constant $US mill)***

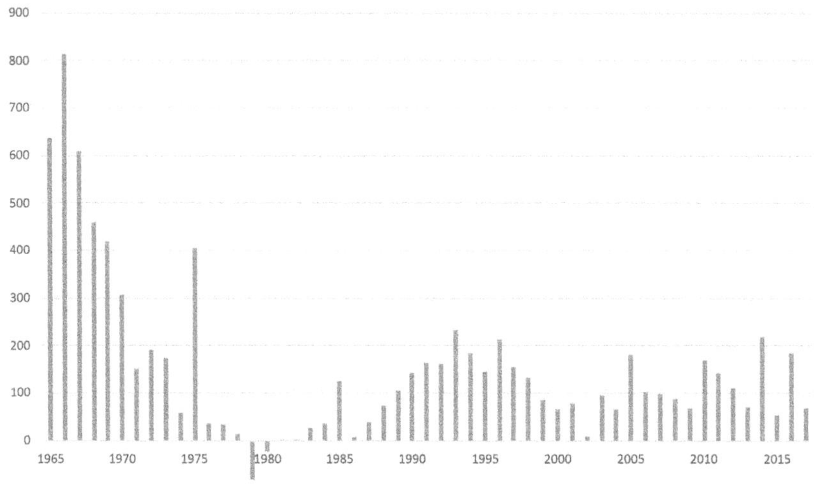

Source: www.stats.oecd.org

Yet, largely due to its rapid economic growth, it applied for membership of the OECD and was admitted in 2010. It is not yet a member of the DAC but it has signalled its intention to become so.

Chile formed its own government aid agency, the *Agencia de Cooperación Internacional de Chile* (AGCI) as far back as 1990 and it has gradually increased its aid budget which it has focused to date on humanitarian assistance, export agriculture and renewable energy in Latin America and the Caribbean among poorer countries such as Haiti, Paraguay and Honduras, for example. Its aid is not yet counted as ODA by the OECD so it is difficult to quantify exactly how much aid it gives. However, an analysis by Gutiérrez and Jaimovich (2017) for the years 2006–2012 suggests that Chile's aid amounts to about 0.01 per cent of GDP which in 2012 amounted to just under $US 30 million, most of which was given to multilateral agencies. For most of this period Chile's ODA donations were equivalent to about 15 per cent of its ODA receipts.

Thus, it seems that Chile is both aid donor and recipient. Aid receipts are directed to a diverse range of programmes, such as renewable energy, vocational training and disaster risk management. As a donor, Chile has targeted not only the United Nations but also a number of bilateral linkages to countries in Latin America. Interestingly its donor and recipient roles have been combined in way recently as Germany as helped fund 'triangular co-operation', whereby projects in Colombia, Guatemala, Honduras, Dominican Republic and Paraguay have used Chile's development experience, funded in part by Germany, to inform programmes in housing, youth unemployment and food security (GIZ n.d.).

All in all, we can see a constantly changing geography of ODA. Looking at the situation in 2019 (Figure 2.2) there is an overall picture of division between the donor world (largely Europe, North America and parts of Oceania and East Asia) and recipients (overwhelmingly still in Africa, South and central Asia, the Pacific Islands, and Latin America). This a conventional North–South

Figure 2.2 ***Current ODA recipients and DAC members***

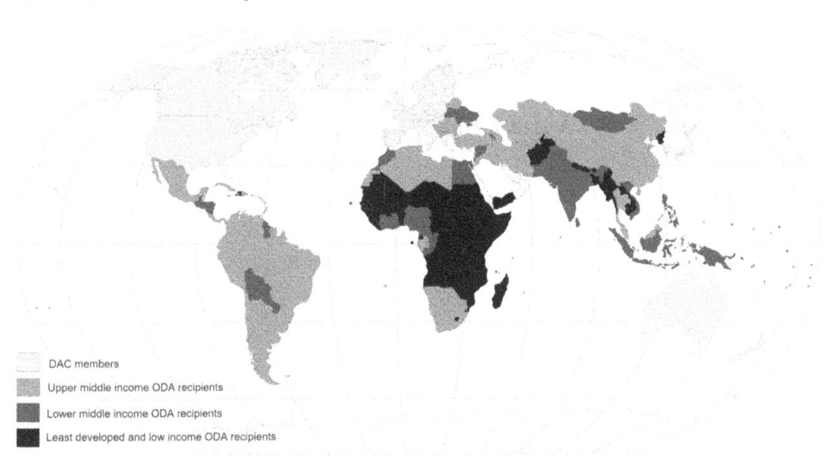

Source: www.stats.oecd.org

model of the aid world. Donors are mostly countries of the 'Global North' and recipients constitute the 'Global South' and this model is embedded in the DAC/OECD system of donors and recipients outlined above. However, changes are occurring with more countries undergoing transition out of recipient status and others straddling both groups.

Types of aid: tied and untied

One feature of aid and ODA is that many of the resources that are provided by donors flow back in some way to them. ODA almost always involve some form of two-way flow and/or elements of reciprocity. These can be quite explicit. 'Tied aid' is that which requires recipients to receive resources that are derived from the donor. These might be, for example, funding for a large rural development project that stipulates that the machinery involved must be purchased from a donor country-based company. Or it might involve technical assistance where the consultants or trainers must come from the donor country. These are ways in which some donors ensure that they get some explicit return to their own economies through the aid they give. Tied aid is a form of subsidy for the domestic economy in the name of foreign assistance. Tied aid also may be less explicit but involve similar two-way flows. The use of aid to support scholarships for students to study overseas in donor schools, technical institutes and universities does train recipient students with important skills (and they are often required to return home immediately after their studies finish) but the money spent to train them (tuition fees, accommodation, living expenses) are spent in the donor economy. Also, in the longer term, graduates return home with personal and professional networks linking them to donors and with skills that may align more with donor systems and ways of working than those at home. Such returns to the donor economy both short- and long-term were once described by a politician in charge of an aid budget as 'doing well from our doing good' (Scheyvens and Overton 1995).

Another growing form of 'aid' – and one that is inextricably tied to the donor country – is the recognition now of costs used to support refugees within donor countries as ODA. Very little of this expenditure is disbursed outside the donor country and although it would seem to meet the welfare needs of a particularly vulnerable segment of the population, it rather stretches the credibility of the way aid and ODA are defined and measured, when virtually all of the expenditure returns to the donor economy. Box 2.4 examines the growth of these costs in ODA accounting in recent years. It has become a significant component of ODA for some countries and, to a large extent, accounts for the growth in ODA in recent years in countries such as Germany. As we will see in Chapter 4, it is also very much in line with a new 'retroliberal' approach to aid, which seeks openly to use aid budgets to benefit donor economies.

Box 2.4 Defining aid: in-country refugee costs as ODA

Costs relating to the resettlement of refugees have been counted as ODA since 1988 but there has been considerable variation in the way DAC countries have reported these. What can be counted as ODA is: 'official sector expenditures for the sustenance of refugees in donor countries during the first twelve months of their stay ... This includes payments for refugees' transport to the host country and temporary sustenance (food, shelter and training)' (DAC Secretariat 2016b). However, this leaves open considerable scope for interpretation. Some donors count the costs of dealing with asylum seekers in their first 12 months. Others include the costs of patrolling, security and the screening of refugees – measures arguably to reduce immigration rather than protect and support refugees (Young-Powell 2017). There has also been a very large increase in such reported costs, especially for European donors in the wake of the Syrian refugee crisis after 2015.

Such increases have attracted scrutiny and criticism from various civil society groups reported a statement from Concord, a confederation of European relief and development NGOs:

> *Supporting refugees arriving in Europe is absolutely the right thing to do, and something we as Europeans should be proud of ... But counting in-donor refugee costs as aid — money spent in the donor country which never reaches a developing country — is of questionable development impact at best and certainly an attempt to artificially inflate countries' aid figures.*
>
> (Lei Ravelo 2017)

Indeed, ODA volumes from several European countries have risen, apparently substantially because of this growth in counting in-country refugee costs as ODA:

> *The biggest increases [in refugee costs as a share of ODA] were seen in Finland (rising from 3 percent in 2015 to 12.3 percent in 2016), Germany (16.8 percent to 25.2 percent), Italy (24.6 percent to 34.3 percent) and Norway (10.8 percent to 18.4 percent) ... However, Austria, Germany, Greece and Italy registered the highest refugee costs as a percentage of ODA spending in 2016, at over 20 percent each.*
>
> (Lei Ravelo 2017)

Globally, such refugee costs appeared to account for about 10.8 per cent of total net ODA in 2016 and 'for the first time, DAC donors spent more on domestic costs ... than on humanitarian assistance' (Young-Powell 2017).

Refugee crises may also be affecting aid in other ways. As well as including in-donor refugee costs, there is increasing interest by donors in using aid overseas to influence migration flows, specifically to slow the rate of migration to Europe and elsewhere. For example, in 2015 the EU established the Emergency Trust Fund for Africa to 'address the root causes of irregular [or undocumented] migration' through the 'flexible, speedy and efficient delivery of support to foster stability and contribute to better migration management' (quoted in Abrahams 2017). By 2017, €2.8 billion had been pledged to the fund. Although money spent to limit migration to a donor country cannot be counted as ODA but if it is used to contribute in some way to development and welfare in a recipient country, then it can. It seems as if many of the projects supported by the Emergency Trust Fund for Africa have indeed been classified as expenditure on 'development' or 'governance' (Abrahams 2017).

Therefore, it seems as if the issues of refugees and migration more generally are muddying our concepts and measurements of aid. Donor concerns about immigration and refugees, so prominent politically in Europe, USA and elsewhere, have influenced the way the considerable costs of receiving refugees and, apparently, preventing immigration have entered into the aid arena. To a large extent, these concerns and expenditures have been largely, if not solely, responsible for the increases in aid volumes in several countries since 2015 and, arguably, responsible for the possible diversion of aid budgets away from development and welfare support in the Global South to activities within donor countries themselves.

'Untied' aid is often regarded as better for the recipients – they can acquire the best and cheapest equipment and advice from the open market and are not forced to spend their aid receipts on what they might consider expensive or unsuitable inputs from the donor. However, even when aid is untied, there are often explicit expectations and/or implicit understandings that recipients will 'return the favour' in some way, be it in terms of diplomatic alignment, political support in international meetings or favourable trade concessions. We will explore these motivations and the way they affect changing approaches to aid in Chapters 3 and 4.

Shifting aidscapes: diverse geographies and types of aid

Yet before we dispense with the concept of aid and ODA, we should recap on what the DAC-centred view of aid has given us.

Despite the recent widening of aid definitions and practices and more explicit donor self-interest, we still face an aid world which has been constructed in particular ways. There is still a dualistic aid architecture marked by a sharp division between supposedly wealthy donors (OECD members) and supposedly needy recipients (on the OECD ODA recipient list). And we see and study 'aid' in terms of the largely state-to-state financial flows that ODA defines and measures.

The increased diversity in approaches does not necessarily spell the end of aid as a unifying concept. Institutionally, and through DAC membership and categorisation, there is a divide between 'donors' and 'recipients'. Some may have a foot in both camps and membership of the categories is continually changing but generally the divide is marked. And geographically, this divide can be depicted starkly. It may not be a simple North–South map (donors in Oceania and East Asia and recipients in the northern hemisphere disrupt that depiction), but it possesses some notable features. There are core donor regions – West and Northern Europe and North America – and there are continuing recipient region concentrations (sub-Saharan Africa, South and Central Asia). However there are now distinct geographies of change: many former members of the Eastern European bloc have transitioned to become donors (Slovenia, Poland, Hungary, Czech Republic); the small island states of the Pacific and Caribbean are still aid recipients (and remain some of the largest per capita aid recipients), but many are progressing to middle- and high-income status; parts of South America (Chile, Uruguay and Brazil in particular) – and perhaps India – are straddling the donor-recipient line; and the rapid economic growth of several Asian states (Singapore, Malaysia, Taiwan, Thailand, China) and several oil-rich states have seen countries become donors, often using their own measures of, and approaches to, development assistance and co-operation.

We are at an interesting historical moment as these trends continue to evolve. The DAC system is still a dominant element of the aid world but it no longer represents an exclusive club of rich donors united by a particular worldview, set of motivations or even geopolitical structure. In this sense we should perhaps talk of different aid worlds – or diverse 'aidscapes'. We still focus on official flows of ODA that we can define and measure and map but increasingly we see other forms of development interaction that are more reciprocal and complex. In order to understand where the aid world might be heading, we now turn to examine forms of 'aid' that have not been captured by DAC definitions and measurements but which have operated significantly in the past and may emerge again in some form in the future. Then we begin to construct an emerging different, and much less dualistic geography of aid.

Re-defining aid: non-ODA forms of assistance

As we have seen, ODA is but one way of defining and measuring the development assistance some countries give to others, though it is undoubtedly the most prominent and widely used approach to analysing aid. Other forms of assistance have been prominent in the past and some persist today. These have been very important forms of 'aid' even if they have not been officially recognised as such and have often defied measurement. Whereas ODA puts stress on financial flows and a stark donor-recipient model of aid, these alternative forms of aid are often more to do with political concessions and legislative changes which do not involve immediate and direct flows of state-to-state resources. Rather they help provide environments which facilitate the ability of individuals and enterprises to operate more freely and globally. They usually lack the strategic and targeted approach of much ODA (to particular projects, sectors and groups) but they nonetheless have been constructed explicitly to provide development benefits. We now turn to examine some of these alternative forms of aid; forms which are more difficult to define and quantify but which may have significant development impacts. Some have been favoured by the major Western donors; others have been pioneered by more non-traditional donors.

Migration

Migration is closely linked to development in a number of ways, yet it is not often regarded as a form of aid. The ability of people in some countries with more restricted economic opportunities and welfare provision to move freely to countries where these prospects are perceived to be better is a major driver not only of human movement but also of a number of development processes.

Firstly, migration can be a way for people to gain access to new labour markets. Relatively better paying jobs overseas, even if accompanied by often very poor working conditions and low wages, are a way for individuals to seek personal gains in wealth and welfare. They may decide to move permanently and gradually sever ties with their country of destination. We can see this as an individual- or family-based development strategy. It is also partly a driver for many thousands of refugees who seek to leave their often conflict-ridden or poverty-affected homes in search of a better life. However, many who move to find jobs in other countries do so in ways which may not be permanent and/or which maintain ties with families at home. In these cases, many of the earnings gained from overseas work are saved and remitted home to relatives. These remittances are very substantial on a global scale and, for some countries, constitute one of the largest segments of national income (Table 2.3). These flows of workers and remittances are hugely diverse. There are domestic workers from the Philippines to be found in Hong Kong, construction workers from Bangladesh and Pakistan in the United Arab Emirates, soldiers from Nepal serving in the British military forces, Mexican agricultural labourers in USA, and nurses from Fiji working in the aged care sector in Australia. Some are able to save and send substantial sums; others may be trapped in work contracts that severely limit their freedom of movement and ability to earn and save above a bare level of subsistence.

Table 2.3 ***Migrant remittance inflows 2018: the top ten by share of GDP***

Country	*Remittance inflow($US mill)*	*Remittances as Share of GDP (%)*
Tonga	165	35.2
Kyrgyz Republic	2960	33.6
Tajikistan	2275	31.0
Haiti	2986	30.7
Nepal	8064	28.0
El Salvador	5458	21.1
Honduras	4746	19.9
Comoros	143	19.1
West Bank and Gaza	2561	17.7
Samoa	142	16.1

Source: World Bank https://www.worldbank.org/en/topic/migrationremittancesdiasporaissues/brief/migration-remittances-data

The issue of remittances has been the subject of debate. Critics (see Gamlen 2014) argue that remittances tend to decline over time as migrants decide to settle overseas and cut their ties; money sent home is often spent on consumables (often imported luxuries) rather than invested in productive enterprises; and labour migration may rob source countries of the young and most able workers, leaving only the old, young or inform as dependents. The latter point is often tied to notions of a 'brain-drain' whereby migration leads to a loss of essential skilled and educated workers (Gibson and Mckenzie 2012). On the other hand, other commentators take a more positive view of migration and remittances, seeing such flows being sustained over time – even if workers do stay overseas they are joined by others and money is still sent home. They also see instances where workers overseas do return home, not only with savings but also new skills (a sort of 'brain gain'), new entrepreneurial attitudes and new international professional and personal networks (UNDP 2009).

Another aspect of migration beyond possible economic benefits are the gains to welfare. Migrants overseas may be able to send their children to local schools or access health care services or even get welfare payments if they are ill or out of work – though such benefits are often constrained by governments for non-citizens. For some, this may be one of the reasons to move: adults can find employment and save their pay packets whilst their children get what they might see as a better education than they could at home. This opens options for others staying more long term – children growing up to become residents and wage earners themselves – or returning home with savings and well-educated children.

These issues with regard to migration may seem to have little to do with aid yet receiving countries can and do make decisions regarding who can enter their countries and these can be done with development benefits explicitly in mind. For example, former colonial powers have often maintained close constitutional relationships with some of their former territories, especially those which are seemingly too small to become viable independent countries. Such relationships are usually maintained because the territories want it that way – they want to retain access to the metropolitan on favourable terms. In the Pacific Islands region, for example, we see the French territory of Wallis and Futuna being recognised as a semi-independent territory, yet its population enjoys the status and rights of French citizens. They can move freely to other French territories, such as New Caledonia, or to metropolitan France and work and gain access to all the educational, health and welfare benefits that French citizens can. Similar arrangements can be seen in the Caribbean, where, for example, the islands of Bonaire, Sint Eustatius and Saba are classified as municipalities of the Netherlands. Thus, in an indirect way, the maintenance of these colonial ties, albeit with a high degree of local autonomy, is a form of 'aid': the right sort of passport can bring a great deal of benefit and higher living standards for individuals, families and whole societies. Furthermore, immigration policy should be seen in a development light: more liberal immigration regimes can be seen as a form of development assistance, whilst a hardening of immigration rules and reduced quota may well have negative development consequences elsewhere.

Another way that migration and remittances may be seen as a form of aid is in emerging forms of managed labour migration. These are separate from general immigration policies, allowing certain people to immigrate if they satisfy specified criteria (skills, age, savings etc), and instead focus on allowing workers to come into a country for a set period of time (usually less than a year but repeat 'tours' may be possible) to work in a particular occupation. Examples of these are found in Australia and New Zealand in the seasonal worker schemes that allow labourers from certain Pacific Islands to work mainly as unskilled or semi-skilled labourers in the agricultural sector (Box 2.5). Here there are shortages of workers – at least at particular times of the year (such as harvests) and for rates of pay at or near the bottom of the labour market.

The seasonal worker schemes are not classified as 'aid' programmes, nor are the costs and benefits counted under the DAC formula for ODA. Yet they are justified as having development benefits for sending countries – they are usually promoted as 'win-win' schemes bringing benefits for the workers and their communities as well as for the agricultural sector in the receiving countries, getting cheap and timely labour supplies. Furthermore, they appear to be popular for the sending countries: political leaders in the Pacific, for example, would like New Zealand and Australia to expand the schemes outlined in Box 2.3, so that more countries and more communities can take part. Even though we cannot easily quantify the benefits and we recognise that 'donors' benefit as much as 'recipients' (and these terms themselves are problematic in this sense), we should see migration concessions and these managed migration schemes as forms of development assistance.

Box 2.5 The Recognised Seasonal Employer scheme in New Zealand: aid through labour mobility?

The Recognised Seasonal Employer (RSE) Scheme allows selected New Zealand employers in viticulture and fruticulture to employ workers from selected countries in the Pacific Islands – Fiji, Kiribati, Nauru, Papua New Guinea, Samoa, Solomon Islands, Tonga, Tuvalu and Vanuatu. Workers are able to stay for 7 months in any given 11 and must show proof of outward-bound passage. Established in 2007 under the Labour government of Helen Clark it has been promoted by aid agencies as a form of development assistance, and one which brings mutual benefits for donor and recipient alike. At its inception 5000 workers per annum were allowed to enter the country under the scheme and this has expanded to over 11000 at the time of writing.

The remittances generated through the work are significant in the economic portfolios of the countries that send labour. A study by the New Zealand government (MBIE 2016) showed that the total combined income for participants after tax in 2014 was just over $NZ110 million and that average individual earnings were above the New Zealand median income by more than 10 per cent. The same study, based on worker surveys, estimated that approximately 40 per cent of this was sent back to home countries, representing a significant proportion of total income in most recipient economies – especially Tonga and Vanuatu where proportional contributions in the order of 20 per cent are estimated. The New Zealand authorities view the scheme in a positive light and are currently studying expanding it to sectors beyond horticulture and viticulture to include other sectors of agriculture and tourism for example. In 2017 it was claimed that approximately 75 per cent of the major wine producing region of Marlborough in South Island was serviced by RSE contractors and this had facilitated a major expansion in exports from this highly successful sector. There were broader positive community impacts as Pacific islanders contribute through institutions such as local churches to local social dynamism (Sachdeva 2017). Those that send workers have also talked of the benefits of the scheme beyond the pure economic returns. The Minister of Employment for Fiji, Jone Usamate, commented that the scheme taught workers from subsistence areas of Fiji invaluable skills that allowed commercialisation as well as contributing to local development through investment in education and housing of the funds (Lewis 2017).

The scheme has not been uncontroversial with criticism aimed at some employers that did not follow the rules fully and some concern with regard to who should have oversight of such issues. Furthermore, private money transfer agencies have also prospered through the system and cheaper forms of remittance transfer have been recommended. As a consequence, the current Labour Government is reviewing the policy with a view to reforms that will deliver greater development benefits for the participant labourers and their communities. There is little doubt that this type of mutual assistance is likely to continue and expand in future years across the world – Australia, for example, introduced a similar scheme in 2008 which is proportionally smaller but shows signs of expanding.

Trade preferences

Another form of assistance, notable in the past, has been the way metropolitan countries have occasionally allowed imports from developing countries on special terms. These concessions operated at a time when, globally, trade barriers were higher than at present and exports from developing countries faced obstacles, such as high tariffs, if they were sent to other countries. Trade protectionism was put in place with the rationale that, by making imports from other countries more restricted and expensive, there would be an incentive for local industries to gain a competitive advantage and thus grow more rapidly. However, larger economies – often former colonial powers – also recognised that such trade protectionism dealt a blow to poorer countries, some of whom might have been former colonies. Others recognised that, even if there had not been an old colonial relationship, there was some sort of moral responsibility to offer concessions to exports from some less well-off countries. In response, systems of trade preferences were instituted in which different schedules of tariffs and/or quotas were used so that favoured exporters faced lower – or no – tariffs or had guaranteed markets allowed for a specified volume of exports.

Perhaps the largest scheme of this kind was that instituted by the European Union with regard to the former colonies of its constituent members. For many years, the European Economic Community[2] used trade protectionism extensively to protect its own agricultural sectors. There were high tariffs and limited quotas on imported agricultural products, especially those which competed directly with European-produced goods. The Lomé Convention was first signed in 1975 and was renegotiated several times so that it spanned the period 1976–99. It allowed for specified volumes of exports from certain African, Caribbean and Pacific countries (ACP) (former French, Belgian, Dutch and British colonies) to enter the European market at prices mostly well above the world average. Products such as beef, sugar and bananas were covered and it meant that the ACP countries party to the agreement had an advantage compared to exporters from elsewhere. It meant that banana producers in Africa, for example, did not have to compete directly with low-cost Central American producers at least in European markets. These were non-reciprocal concessions – the ACP countries did not have to offer favourable trade terms to European exporters in return. The Lomé Convention undoubtedly provided a significant form of development assistance for those lucky enough to be a part of the agreement. Producers not only had a secure market to sell to at favourable prices but also those prices were reasonably predictable, thanks to the institution of stabilisation funds which ironed-out major price fluctuations. In effect, 'aid' was given by European consumers, in the form of higher retail prices paid, to producers in former colonies.

There were other preferential trade schemes. The USA, for example, recognised a schedule of preferences for the producers of certain manufactured items, again with a guaranteed quota of imports for particular products from specified countries. Although these quota schemes were gradually abolished under World Trade Organization (WTO) agreements in the early 2000s, some products and countries, such as garments from Nepal or Cambodia, can still be imported duty free into USA. As with the Lomé Convention, these trade preferences have allowed metropolitan powers to exert influence over their partners – the threat of withdrawal of benefits would have been a huge economic blow for a country falling foul of their patron. Yet there were many examples of success. Fiji is an interesting example. This Pacific Island nation of about 800,000 people faces difficulties in competing on global markets, due to its relatively small size (not allowing for large economies of scale) and distance from world markets. However, under Lomé it was able to preserve and expand its sugar industry, established during colonial times, and this formed the backbone of the economy, employing farmers, contractors and mill workers and supporting a relatively vibrant rural economy. In addition, trade preferences were later allowed by USA, Australia and New Zealand so that Fiji-made garments could be imported to these countries at low tariffs (compared to imports from, say, Thailand or Indonesia). The garment industry thrived for a time in the 1980s and 1990s and provided jobs (albeit low-paid work) for many and the industry briefly challenged sugar and tourism as the leading economic earner in the country.

This system has been termed 'aid with dignity' (Taylor 1987). It allowed developing world economies to focus on producing needed commodities and for these they received a good steady income. There was no hand-out of aid money, no ODA to measure; rather there was a recognition of special historical ties and responsibilities in ways which supported the economies of developing countries.

However the Lomé system and other preferential schemes could not outlast growing global trade liberalisation that became part of neoliberalism in the 1990s. Particularly as a result of US pressure to remove any trade preferences, the Lomé Convention ended in 1999 to be replaced by the Cotonou Agreement which did not have the same guarantees or security for ACP exporters and required more reciprocal concessions. Now with a much more open global trading environment, where low-cost producers exploit economies of scale and low wages, many of these former industries have collapsed. Smaller economies cannot compete; others face new forms of trade barriers (relating to biosecurity etc.); and many are forced to accept trade agreements that not only involve a loss of guaranteed markets for their exports but also the end of forms of protection for their own industries. In this sense, the loss of trade preferences as an implicit form of aid before 2000, should be seen alongside the substantial increases in ODA thereafter. A hidden form of aid to promote economic growth and employment was replaced by another more explicit form and one where the power imbalances became more marked. 'Favoured trading partners' became 'ODA recipients' and level of assistance in a holistic sense did not necessarily rise.

Loans

We have already seen that concessionary loans are counted as ODA. In other words, loans are considered a form of ODA if they are offered at terms and rates more favourable than world market rates. What is counted as ODA is not the amount of the loan principal but the difference in interest rates between the reigning market rate and the concessional rate offered (as long as it meets the DAC's 25 per cent grant element equivalent standard). This takes our understanding of aid into new realms because it is possible that, over its lifetime, an aid loan will result in a net flow of resources from the recipient to the donor (repayment of principle plus interest even allowing for inflation) rather than the other way around. The 'aid' is merely a discount offered on a financial transaction. Furthermore, these concessions occur in a global financial system where all parties are far from equal. A well-developed and stable industrial economy may have an excellent credit rating and be able to secure loans from financial institutions at relatively low rates of interest (because it has a low risk of failure). Conversely, a developing country with relative instability and poor economic performance will face a bad credit rating and high interest rates. In other words, precisely those countries that need loans for development have to pay a lot more than well-off industrialised economies. For many developing economies, obtaining loans on the open financial market for, say, a large infrastructure project or a hydroelectric dam, is both difficult and expensive.

Aid in the form of concessionary loans, can therefore be a critical source of development finance for poorer economies. International financial institutions, major development banks in particular, play important roles here. Development Banks, such as the World Bank, the Asian Development Bank, BancoSur or the African Development Bank, operate in an interesting way. They are backed by large donor economies (USA, Japan, Germany, France, UK etc.). These donors are able to use their own good credit rating to raise (and guarantee) loans and then pass these favourable terms on to selected recipients, who meet the requirements of the bank concerned. Projects are closely vetted and, in theory, their development benefits and ability to generate revenue to repay the loans are laid out before the loan is approved. Such loans, by the World Bank for example (Box 2.6), have been highly significant sources of finance for development projects across the developing world. In most cases, although they were relatively tightly planned and monitored, they did allow borrowers to use the funds to purchase inputs on the open market. 'Multilateral' agencies (with many donors and many recipients) seemed to escape the 'tied aid' problem.

Box 2.6 Key institutions: the World Bank

The World Bank, together with the International Monetary Fund (IMF), was established following the Bretton Woods Conference of 1944 as institutions and mechanisms to stabilise and regulate the world economy following the Second World War. The Bank focused first on the reconstruction of European industrial economies decimated by the war, but then turned in the 1960s to focus more on supporting the development of newly independent countries in the Global South.

The Bank is organised into two key units, the International Bank for Reconstruction and Development (IBRD) and the International Development Association (IDA) and also encompasses the International Finance Corporation (IFC) and some smaller agencies. It is governed by a Board of Directors and its President by convention is always a US appointee. Historically, voting power has been dominated by the main donors to the Bank, principally the USA, Japan, Germany, UK and France, but since 2010 this has broadened somewhat to include China, Russia, India and Saudi Arabia. The dominance of the main Western donors has led to claims that this has led the Bank to evolve policies which ostensibly are designed to foster development but in reality are intended to maintain the economic dominance of the donors.

The World Bank works mainly by providing relatively low interest loans for governments. It uses the credit worthiness of its main donors to raise funds which are then, in effect, on-loan to emerging economies that would otherwise face higher interest rates on commercial financial markets, due to their higher risk assessments. Loans are given out with strict conditions, and these have evolved over time. During the 1980s and 1990s in response to the debt crisis – which in part was facilitated by the lending of the financial institutions under more positive economic conditions in the pre-oil crisis world – strict structural adjustment programmes had to be adhered to. These included privatisation, deregulation, export orientation and other policies associated with neoliberal economic policy. This led to a 'lost decade' of development in the Global South which the World Bank saw as part of a necessary adjustment to harsh financial realities of globalisation. In the face of criticism regarding such polices the emphasis was shifted to poverty reduction strategies in the late 1990s and 2000s, although it could be argued that similar free-market policies that favour global corporations were fostered whilst masquerading as socially progressive agreements. In the 2000s also debt cancellation policies were adopted for the most heavily indebted countries (approximately 40, many in Africa).

Since 2000, and since criticism of its structural adjustment policies of the 1990s, the Bank's policies have been explicitly linked to a poverty alleviation mandate, first defined by the MDGs and, since 2015, by the SDGs. There has also been a recognition of the need for the Bank to help address climate change. Yet, when we look at where the Bank's lending goes, we see that loans for infrastructural development (roads, railways, electricity, irrigation etc.) still dominate, and there has been increasing attention paid to government decentralisation. Welfare services, health and sanitation are also supported but they are not as prominent as might be expected given the poverty alleviation rhetoric.

Not all funded projects succeeded by any means, and there has been widespread criticism of some of the lending policies of the World Bank (see Box 2.6) and others in the past (see Toye *et al.* 2013). These have concerned not only the ability of governments to repay the loans but also factors such as harmful environmental and social impacts associated with the free-market neoliberal-inspired restructuring upon which such loans are conditional.

Loans from the World Bank and other institutions are counted as ODA as long as they satisfy the 25 per cent concessional element, as are the frequent grants they make (funds given without a requirement to repay) often alongside the loans. However, we are beginning to see a new global financial landscape appearing where new lenders outside the OECD/DAC framework are appearing. These are institutions whose lending is sometimes not as transparent as their established counterparts – terms and conditions are shielded because of their economic or political sensitivities. In addition, many of these new development concessional loans are being offered by a single donor – China in particular – and they are very much of the 'tied aid' form: materials, equipment, technical expertise, even labour, must be purchased from the donor.

However, an interesting initiative is the 'New Development Bank' which began operations in 2016 and has its headquarters in Shanghai. Its five founding members, all of whom contributed $US10 billion each to the Bank's authorised capital, are the BRICS countries (Brazil, Russia, India, China, and South Africa). This bank will focus on infrastructural and energy projects and is likely to have a significant impact on development finance globally, augmenting the conventional sources such as the Asian Development Bank (ADB) and World Bank. Initially it is likely that most lending will be to projects within the BRICS countries but this will spread beyond. Quite how the concessional elements of the Bank will operate – and how or if they can be calculated as 'aid' – are not clear. However, what we are seeing is a major new aid organisation appearing and working largely outside of the old OECD 'club'.

Welfare assistance

As with these new lending institutions, there are other forms of aid that are expanding and appearing outside the conventional North–South model. We have seen that the DAC system captures aid for welfare as ODA. This includes the very large sums spent on health services or education in many forms as well as a multitude of projects and programmes focusing on issues, such as gender-based violence, women's empowerment, community development etc. These all amount to a significant portion of the global flows of ODA (see Chapter 3).

However, not all such forms of welfare assistance can be tracked through the ODA system. We do not consider here the unofficial forms of welfare assistance carried out by individuals and families (relatives paying and caring for children attending schools and universities overseas, or family members being sent to receive health care in another country). Such flows would be nearly impossible to calculate. Yet we do recognise that some important forms of welfare support do exist and are not adequately recognised or measured.

There is also the issue of the work of civil society organisations. NGOs such as Child Fund and World Vision and a host of others raise funds in the West and provide health and education services elsewhere. Some of this might be captured in official statistics; much is not. The phenomenon of child sponsorship, for example, is responsible for large sums being transferred across the globe (Wydick *et al.* 2013).

Emerging aidscapes: South–South co-operation

But it is with some of the non-traditional (non-DAC) donors that some of the major flows of aid can be seen (at least in part). The granting of scholarships by donors to students from the developing world to attend institutions in the donor countries has long been a popular form of aid. Selected students get a chance to broaden their education and life experiences overseas whilst donors also benefit by establishing networks of understanding and relationships that persist beyond the time at school or university. These scholarships from DAC donors are counted. However, new donors see similar benefits in scholarships. Having students from another country study in the donor's own institutions builds and strengthens ties and open up channels for future collaboration. Language and cultural barriers are weakened as more people are able to operate in two or more worlds. China, in particular, has seen great merit in such scholarship schemes and has signalled major expansion of this form of support.

Box 2.7 Measuring Chinese aid

The PRC has emerged as a major aid donor on a global scale in the past two decades. Although its aid is not counted as ODA by the OECD/DAC and China itself refers to its gifts and loans as forms of 'co-operation' rather than 'aid', nonetheless the growth and very large volumes of its loans and grants to developing countries have seen it recognised as one of the world's largest donors – and the source of some suspicion and even envy from more established donors. Its own development strategies, such as the 'One Belt, One Road' initiative, involve building much stronger economic and other relationships with developing countries in Asia, Oceania, Africa and South America. It is attempting to promote a new form of globalisation; one that sees China more at the centre, rather than the existing USA-European core. Infrastructural development is vital for constructing this new economic world and here China has been active in assisting countries to build new ports, roads, telecommunications systems, and energy facilities.

Yet, if we cannot define these forms of assistance – many in the form of loans (with opaque terms and conditions) tied to contracts let to Chinese contractors – as ODA and we lack both comprehensive and accurate data on one hand and criteria for classifying these as 'aid', then how can we measure how much 'aid' China gives? Chinese aid and investment – and that of several other countries such as Saudi Arabia and India – is not tracked through the OECD/DAC system and is therefore seen to lack transparency. However, if we see these flows as commercial transactions and 'development financing' rather than ODA, then China is perhaps no different from large global corporations or many countries in not publishing the details of all its financial transactions and commercial deals.

Nonetheless, attempts have been made to estimate the volume of Chinese aid. Devex (2013) produced some broad figures that suggested that what could be considered Chinese ODA was about $US 2 billion at a time when US ODA stood at $US 30 billion. However, it was also noted that China's 'other financial flows', particularly export credits, were significantly larger than ODA at between five and six billion dollars, far in excess of USA. So, on the basis of this, China might be seen as a very significant ODA donor, but not amongst the world's largest. Yet its wider operations involving net flows to other countries would put it in this category, alongside France, Germany, UK and Japan, if not USA.

In an interesting paper, Caitlyn Sears (2019), following the earlier work of Bräutigam (2011) and others, has attempted to disentangle Chinese aid and apply some frameworks to measure the volume of aid. She pointed to several different ways of thinking about and quantifying aid that China disburses and which OECD/DAC does not recognise adequately. This includes in-kind forms of assistance and nonmonetary aid, such as the sending of experts to advise on technical matters or the building of hospitals or sports facilities (where these are not covered by formal contracts). She also noted that the use of loans on interest-free or concessional (and flexible) terms presents another problem of quantification next to the OECD/DAC 'rules' on measuring these.

Looking at the complexity of the data available, she produced a rather better estimate compared to the Devex exercise above. She concluded that in 2010, China disbursed about US$2.8 billion in grants, US$125 million in interest-free loans, and US$6.2 billion in concessionary loans (Sears 2019: 141). Again, though, these data place China as a significant aid donor, but not the largest. In Chapter 3 we will look in more detail at how Chinese aid operates in the Pacific region.

Similarly, medical training is a way of building good diplomatic relationships by utilising the donor's own expertise and institutions. In this regard, the example of Cuba is notable (Box 2.8). By many indicators, Cuba would be classified as a low-income country, apparently in need of development assistance. Yet since the revolution of 1959 it has taken particular pride in prioritising the health and education of its own citizens. It has invested heavily in these services, arguably at the expense of other sectors, but its education and health indicators stand up extremely favourably by international standards. As part of this investment, Cuba has developed a particular model of medical training that is different from most Western models. Eschewing high-tech medical models which see heavy investment in large hospitals, expensive equipment and high-end medical training, Cuba's model focuses instead on primary health and in the provision of well-trained professionals serving the needs of all communities. It is a model that seems to suit cash-strapped developing countries and Cuba has been active in offering not only its own doctors to serve overseas, but also to open its training institutions to students from overseas (Box 2.8).

Box 2.8 South–South co-operation: Cuban medical aid

Following the revolution of 1959, Cuba has pursued the policy of building solidarity and relationships through bilateral health alliances. A central component of this is the sending abroad of trainees and recently qualified young medics from Cuba to poorer countries throughout Latin America, Asia, the Pacific and in particular Africa. Cuba has invested heavily in the health sector and long seen generous provision as a central pillar of its socialist path. A particular objective of Cuban health policy has been to ensure that all citizens have access to primary health care. Health professionals are found in nearly all communities and there is emphasis on health education and low-tech health treatment. This is in contrast to Western medical models that have more of an emphasis on secondary and tertiary health and higher-cost health facilities. Where primary health care is a priority, the skills and abilities of Cuban doctors are well-respected and sought after in countries that have been unable to develop such capacity.

In 2012 Cuba had over 38,000 doctors placed in 74 countries (Asante *et al.* 2012). When Cuban doctors return home to Cuba they are held up as returning heroes of the revolution spreading skill and compassion across the world. The medical assistance programme is framed more as a matter of 'solidarity' with people overseas than 'aid'.

Increasingly, Cuban medical aid has turned to training doctors from other countries rather than sending Cuban doctors overseas. In 2012, there were some 20,000 medical trainees from over 60 countries studying in Cuba (Asante *et al.* 2012). Again, the training they receive is not the same they would get in Western medical schools but the lower costs of training and the emphasis on primary health care mean that many more medics can be trained and have a great impact on the health needs of many more people, including those in more remote areas. It is regarded as a more sustainable health care model by many, given the resource constraints of poorer countries and the very high costs, and difficulty of retaining, Western-trained doctors. In Timor-Leste, for example, Cuban-trained medical staff account for nearly three-quarters of the country's doctors (Asante *et al.* 2014: 277). Initially, most East Timorese trainee doctors studied in Cuba. Up until 2007 some 677 enrolled for the medical degrees in Cuba but thereafter, a steady stream (a total of 328 between 2005 and 2011) enrolled for their training in-country, through Cuban instructors (the Cuban Medical Brigade) and the local university. Again, many of these graduates serve in rural areas. In many ways, this has represented a remarkable programme of assistance that has provided the new country with a whole generation of medics, a health care sector that would have not had nearly the same numbers, nor the same reach, had more mainstream models of training and aid been employed.

There have been criticisms of this Cuban medical assistance model, including language barriers (doctors are trained in Spanish and this may conflict with local vernacular and English-dominated facilities), the lack of high-end medical skills and the inability of recipient countries to employ those medics that are trained. However, there can be little doubt that overall the system has positive repercussions over time and represents an alternative to the Western-led orthodox model of aid delivery.

The examples of Chinese educational scholarships and Cuban medical training, then, are interesting as they show how aid is moving in new and innovative directions. Neither example is free of criticism and some charge that it is not synchronised with local systems, languages and practices. However, it is likely they will continue to expand and we may see other examples, such as scholarships to study in India, appear in future.

In this context, we can begin to understand more generally how 'aid' is being reconstructed, less as a dual Northern donor/Southern recipient binary model, and more as a complex web of economic relationships linking countries of the South in systems of co-operation and joint development. It is offering a major challenge to the way we think about aid and indeed to the very concept of development (Kim and Lightfoot 2011, Kilby 2018).

As noted briefly in Chapter 1, it has been suggested that we are seeing a new aid world – or even a 'post-aid world' emerging (Mawdsley *et al.* 2014). A key part of this new aid landscape is what Mawdsley terms the 'Southernisation' of aid: 'A more polycentric global development landscape has emerged over the past decade or so, rupturing the formerly dominant North–South axis of power and knowledge' (Mawdsley 2018: 173). Previously, the DAC-centred aid world was predicated on the power of the main Western donors and their attempts to 'socialise' emerging economies, such as Korea, so that they would behave like the established donors, eventually joining the DAC and abiding by its conventions. However, China especially, but also other growing economies, such as India and Brazil, have chosen to operate rather differently and largely outside of the DAC framework (Mawdsley 2010, 2012a, 2012b).

They have pursued new practices and languages: 'South–South co-operation' has adopted the mantra of win-win development, whereby new economic and political relationships are forged. These may involve net flows of both financial and nonmonetary resources from one country to another (what we might see as 'aid') but aid is not the primary basis of the relationship. China is open about how these relationships are for its own geopolitical and economic benefit, as well as for its partners. It is a different ethos of assistance, moving from one based on supposed altruism to one built upon a vision of explicit shared benefit.

Loans (concessionary and otherwise), scholarship schemes, business deals, investment and gifts all cement relationships which have the overt goal of mutual benefits. And this challenges the power relations implicit in the DAC-centred aid model. Wealthy and powerful donors giving to supposedly poor, weak and needy recipients are replaced by a web of reciprocal and complex trade, investment and technological interactions which appear to be amongst partners of equal standing, even if not of equal wealth.

These are significant and growing challenges to the established aid world centred on Washington or Paris. We argue in Chapter 4 that Western donors have reacted to this South–South challenge by adopting their own 'retroliberal' approaches to aid, eschewing the strong poverty-focus of the earlier neostructural regime and instead promoting narratives of 'shared prosperity', sustainable economic growth', and unashamed donor self-interest. Yet we also add a note of caution. We would answer Kilby's (2018) questioning whether 'DAC is dead' in the negative. Despite the undoubted Southernisation of aid, despite the rise of China, despite the retroliberal turn and despite the growing complexity and obfuscation of aid, the DAC system is still very much in place, the narratives of 'aid' and 'poverty' are still strongly adhered to, and the flows of resources from the Global North to the South in the form of ODA are still very substantial and showing little sign of ending.

Conclusions

We began this chapter by suggesting that 'aid' was a relatively simple concept, to do with 'assistance' and 'help' and implying that better-off countries aid those less well-off. In relation to international aid we then saw how there exists a relatively precise definition in the form of ODA which is widely accepted and measured internationally. Aid as ODA requires assistance to be spent on development and improving welfare, it should flow through official channels and it must involve a net benefit for the recipients.

However, as we broadened our discussion, we found that these fairly precise meanings are not so clear in reality and present many difficulties in terms of their measurement. The interpretation and quantifying of ODA has experienced some changes over time and there are many forms of assistance, such as trade and migration concessions, that are not counted. We might also suggest that the notable increases in ODA since 2000 have come at a time when the other forms of assistance, especially trade preferences, have been eroded. Yet despite these complexities of definition and the need to keep a watch on 'non-ODA' forms of aid, ODA remains the most useful and precise concept for us to use to analyse key aspects of international development assistance.

Similarly, the parties involved in aid have altered over time. When we adopt the DAC framework and just look at ODA, there is a reasonably clear global landscape of aid with donors largely being the developed countries of the Global North (North America, Western Europe, Japan and Australasia – all members of the OECD) and recipients being located largely in the Global South (Africa, Asia, Latin America and Oceania). Yet this geography is being transformed in different ways. A number of countries have undergone a transition from low- to middle- or high-income status – Korea, Singapore, Chile and Saudi Arabia for example. Some have joined the OECD/DAC 'club' and become donors themselves. Some, India included, have become a recipient and a donor simultaneously. Others again, such as China, have forged their own path, also becoming substantial donors but adopting their own ways of operating. These forms, including concessionary loans, grants and forms of welfare assistance lie outside the DAC framework. Increasingly, then, the old North–South dichotomy is no longer relevant. In later chapters we will see how new forms of South–South co-operation and the emergence of non-traditional donors are lining up alongside the changing scene within the OECD aid framework.

Therefore, what we have seen is that aid is not only complex, it is rapidly changing. As Sears suggests, 'aid is not a static measure but rather a moving target' (2019: 141). We need to adopt a broad view of what we mean by 'aid', who we see being involved, and how aid relationships and flows work in practice. In the chapters that follow we start by examining the detailed geography of aid, focusing mainly on ODA and seeing where aid comes from and where it goes.

Summary

- Aid is a broad concept and must include all sorts of 'assistance' as well as policies that are designed for mutual benefits for donors and recipients.
- Aid worldviews have shifted significantly over time from those based on relief, to those targeting development, advocacy, self-interest, and latterly private sector development.
- These aid worldviews have evolved chronologically to an extent, but there are areas of policy and conceptual overlap.
- We may be entering a post-aid world, based on 'shared prosperity' and 'South–South' co-operation among other things. The motivations for aid have become highly complex and difficult to pick apart to the extent that terming it 'aid' at all is contestable.
- The official OECD DAC definition of aid has changed over time to become more inclusive. However, it still fails to capture many types of aid that are not measured, such as migration and labour schemes, trade preferences, concessional loans, and welfare assistance.

- On the other hand, ODA has come to include some financial flows, which are mostly or fully retained in donor economies, such as in-country refugee costs.
- A number of countries, such as China, are not in the DAC, yet their aid activities are large and growing.
- There is increasingly a blurred line between donors and recipients as development patterns shift and geopolitics change. The resultant geography of flows is highly variable and complex.
- We discussed both tied and untied forms of aid, noting that a shift away from the former has now been, at least in part reversed.
- There has been a large shift in the Global South towards what some have called 'South–South' co-operation with scholarships, training schemes and other forms of mutual assistance patterns. China has been especially prominent in evolving such models.
- Notwithstanding the 'southernisation' of aid we argue that the traditional DAC form of ODA remains the most important of all and that we do not yet live in a post-aid world.

Discussion questions

- Discuss the evolution of aid worldviews over time and the extent to which these worldviews overlap.
- What aspects of 'aid' are not captured in the official OECD DAC definition and what are the implications of this for understanding the patterns and impacts of aid?
- What aspects of ODA are included in its definition, but might be questioned as 'aid' from donors to the Global South?
- What do analysts and observers mean by South–South co-operation and to what extent is it replacing more traditional forms of ODA?
- To what extent is the hypothesis of a 'post-aid' world a reasonable assessment?

Websites

- World Bank: https://www.worldbank.org/en/who-we-are
- OECD/DAC: http://www.oecd.org/dac/development-assistance-committee/
- Devex: https://www.devex.com/news
- Lowy Institute (Australia) on China: https://www.lowyinstitute.org/issues/china

Notes

1 In DAC statistics, this implies a grant element of at least …

> • *45% in the case of bilateral loans to the official sector of LDCs and other LICs (calculated at a rate of discount of 9 per cent).*
>
> • *15% in the case of bilateral loans to the official sector of LMICs (calculated at a rate of discount of 7 per cent).*
>
> • *10% in the case of bilateral loans to the official sector of UMICs (calculated at a rate of discount of 6 per cent).*
>
> • *10% in the case of loans to multilateral institutions (see note 5) (calculated at a rate of discount of 5 per cent for global institutions and multilateral development banks, and 6 per cent for other organisations, including sub-regional organisations).*
>
> (OECD 2019b)

2 The European Economic Community was established in 1957. It grew in membership from the original six and in 1993 became the European Union, which in 2002 enacted a currency union for the majority of the members. There are currently 27 members. The EU deals with common economic, social, and environmental policy and as a multilateral body has an important role on the international stage – including in the aid world.

Further reading

Alfini, N. and Chambers, R. (2007) 'Words count: Taking a count of the changing language of British aid', *Development in Practice* 17(4–5), 492–504.

Kilby, P. (2018) 'DAC is dead? Implications for teaching development studies', *Asia Pacific Viewpoint* 59(2), 226–234.

Kumar, R. (2019) *The Business of Changing the World: How Billionaires, Tech Disrupters, and Social Entrepreneurs Are Transforming the Global Aid Industry*. Beacon Press, Boston.

Lancaster, C. (2007) *Foreign Aid: Diplomacy, Development, Domestic Politics*. University of Chicago Press, Chicago.

Mawdsley, E. (2018) 'The 'Southernisation' of development?', *Asia Pacific Viewpoint* 59(2), 173–185.

Swiss, L. (2016) 'World society and the global foreign aid network', *Sociology of Development, 2*(4), 342–374.

Toye, J., Harrigan, J. and Mosley, P. (2013) *Aid and Power-Vol 1: The World Bank and Policy Based Lending*. Routledge, London and New York.

3 Patterns of aid

Learning objectives

This chapter will help readers to:

- Understand global flows of aid over time and space
- Describe the relative importance of different aid donors and recipients in the contemporary world
- Explain the difference and similarities between traditional and non-traditional aid donors
- Account for the rise of new donors, and appreciate what is motivating this
- Appreciate the wide range of aid targets between different donors
- Understand the role of historical ties and geopolitical factors in determining patterns of aid flows
- Debate whether aid is allocated according to need or other factors
- Distinguish between different types of aid recipients in terms of what it is that motivates the flows to them
- Understand the recent widening of the range of aid recipients and account for this change

Although aid is difficult to define and measure as we discovered in Chapter 2, it is possible to ascertain and describe some of the major flows of aid resources from country to country and from agency to agency. In this chapter we attempt to map some of those major flows, looking at both absolute (total volume) and relative (share of Gross National Income (GNI) and per capita) measurements. We also unpick some elements of the types of aid that move between donors and recipients. We focus particularly, though not exclusively, on official (Official Development Assistance (ODA)) data and the mainstream aid landscape.

A geography of aid

The patterns of aid flows are not straightforward. In the above chapter, we saw that agencies change over time and not all forms of aid can be easily quantified and compared. Figure 3.1 depicts a framework for seeing where aid comes from and where it flows to. Firstly, there are the primary sources on the top line. These come mainly from governments, whether members of Development Assistance Committee (DAC) (USA, Japan, UK, France etc.), present or aspiring members of the OECD, but not yet DAC members (Chile, Estonia, Russia, Costa Rica), or non-OECD countries (China (People's Republic of China – PRC), Chinese Taipei/Taiwan (Republic of China – RoC), Saudi Arabia). Donors can also include a category of countries that are simultaneously donors and recipients (India). Other primary donors include private funders (the Bill and Melinda Gates Foundation is notable in this regard) and a plethora of non-government organisations (NGOs) who raise money themselves from the public and governments.

Figure 3.1 ***Schematic diagram of ODA flows***

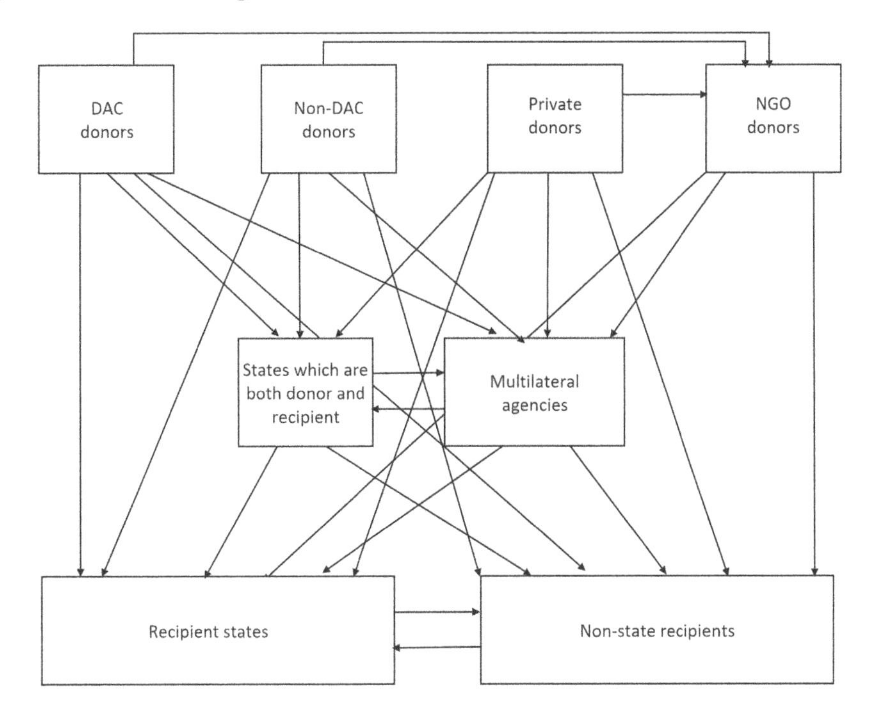

Below these primary donors are the very significant multilateral organisations (the UN, a range of development banks, and official organisations with specific missions, such as the Global Environmental Facility – GEF). These organisations derive their income from donor states as well as, in the case of development banks, from their own trading operations and sometimes from NGOs and private sources. The share of total ODA flows over the five year period 2013–17 is depicted in Figure 3.2. We see that DAC donors account for over half of all ODA but what is perhaps surprising is the large share – about one-third – from non-DAC countries. This is probably the result of the very high ODA returns from Saudi Arabia, Turkey and UAE in this period (see Figure 3.7), much related to the flow of refugees from Syria after 2015 and humanitarian crises in Myanmar and Yemen.

Flows of aid from donor organisations follow a variety of paths (Figure 3.1). There are *bilateral flows* – perhaps the most common and largest of the flows – and these link a donor with a recipient

Figure 3.2 ***Share of global ODA by donor category 2013–17***

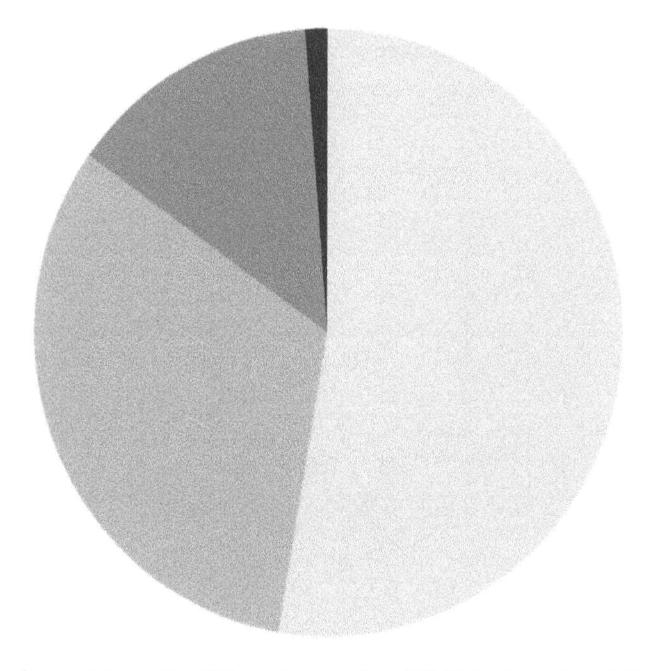

Source: www.stats.oecd.org

country in a series of one-to-one relationships (hence 'bilateral'). There are the *multilateral* flows: these see several sources of funding flow into an organisation, which then disperses it to many recipients. Both bilateral and multilateral flows, especially when involving OECD members, tend to be captured fairly fully as ODA statistics, though many of the funds channelled between non-OECD members and recipients are not. Some private and NGO flows (largely from large organisations) are counted by DAC and these mainly involve flows from the organisations to non-state projects and programmes, though some funders (the Bill and Melinda Gates Foundation and some of the larger NGOs) will support certain projects by multilateral and state agencies from time to time. Mostly, however, the NGO sector works outside of the ODA environment and their aid flows are not analysed, except for when donor states contribute to the work of NGOs and fund them directly. Donor governments also fund projects directly in recipient countries, outside of state agencies, especially when recipient states lack the capacity or legitimacy to perform certain development functions.

In order to analyse these flows, we mainly make use of DAC ODA statistics. These, of course, do not measure all aid flows, but the ODA volumes they do record give a good indication of relative size and direction. Here we use data on 'aid disbursements', those funds actually spent by donors (rather than just budgeted), though they might not be a completely accurate record of what funds finally make it through recipient agency systems to the desired end-point. In order to account for fluctuations from one year to another, for example the occurrence of a major disaster or a short-term economic variation, we use a five year average over the years 2013–17 and we allow for inflation by using the adjusted 'constant' values (at 2017 values for the $US) rather than the actual 'current value' in each year. There are problems with any such measures, such as using the US dollar as the standard, but again these results give a reasonable view of overall ODA patterns.

Traditional and established donors

Established donors are those who work within the DAC framework and who account for the largest share of ODA flows (see Figure 3.2). It is the aid budgets of these countries that fund the majority of the bilateral programmes worldwide and also support the work of key multilateral agencies. To a large extent these established donors are the original large founders of the DAC: USA, UK, France, Germany, Japan, and the northern European countries of Sweden, Norway and the Netherlands. It is striking just how much of global ODA comes from a small number of countries. The USA, for example, alone accounts for 30 per cent of the total given by DAC country members in 2013–17 and the largest five donors (USA, UK, Germany, France and Japan) account for 69 per cent (see Figure 3.3). Even if we account for all ODA donors including multilateral institutions, non-OECD donors and private sources, the figures are still impressive: USA contributes 20 per cent of global ODA and the five largest donors 43 per cent.

Whilst the 'big five' dominate total volumes, the contribution from other states is significant. Another ten countries average over $US 1 billion per year in ODA and the DAC list is continually expanding, bringing many of the former transition economies of

Figure 3.3 ***DAC donors ODA annual average 2013–17 ($US mill constant 2017)***

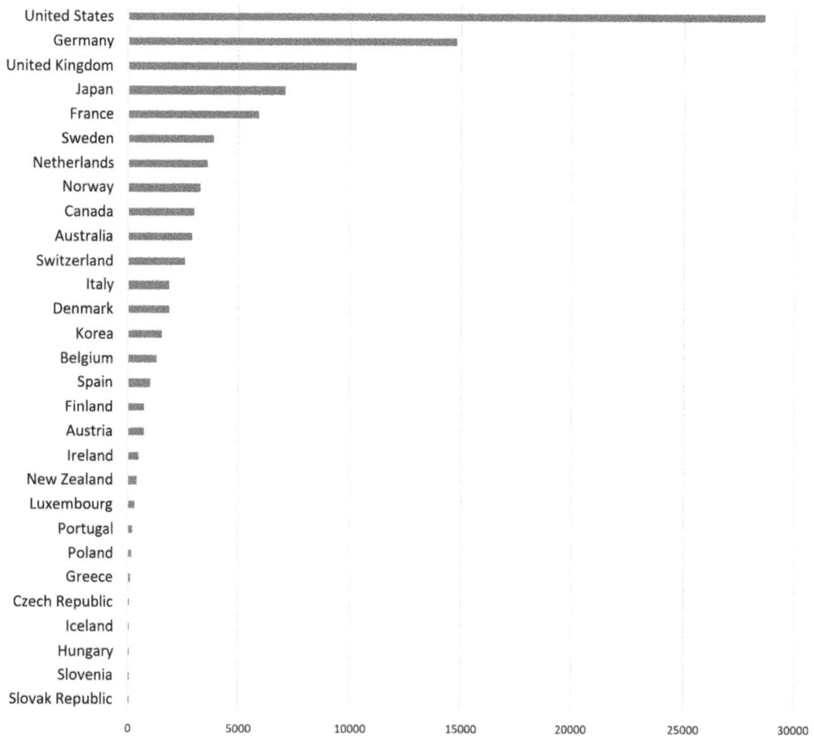

Source: www.stats.oecd.org

Eastern European (Poland, Slovenia, Hungary etc.) and still includes countries which experienced severe economic problems themselves in the past decade (Greece, Portugal and Spain).

Another way of analysing aid donations is to look at the relative 'generosity' of donor countries. USA and the other major donors dominate total volumes because they are amongst the world's largest economies (only China would be able to match them in terms of economic power). When we measure ODA against the size of the national economy (GNI) we get a rather different picture (Figure 3.4). In this case it is the countries of northern Europe (Sweden, Norway, Luxembourg, Denmark, Netherlands and UK) that stand out as well as the UAE. Indeed, only these seven countries have managed to match or exceed the 0.7 per cent of national income target set in 1970.[1] Further down the list, European countries continue to dominate.[2] Furthermore, it is apparent that the large USA and Japanese total volumes recorded are not matched by their proportionate commitments to ODA: both hover at or below the 0.2 per cent of GNI mark.

Finally, with regard to the established DAC donors, it is possible to see some interesting changes over time. Although we will look at the way aid policies and motivations have evolved over time in detail in Chapter 4 and we have briefly reviewed these at the start of Chapter 2, Figure 3.5 shows that the established donors have always dominated ODA volumes but their relative positions have altered since 1970. Again we use dollar values adjusted for inflation (constant $) to see changes in 'real' ODA values over time. Only selected major donors are shown to indicate some broad features.

Firstly, it is apparent that USA has always been the largest ODA donor, though the value of ODA has changed markedly over the years. Some fluctuation was apparent between 1970 and 1990 but the 1990s saw a sharp decline in real values before a sharp and steady rise through the first decade of the new millennium. Since 2010 there are perhaps signs of a decline. These trends were matched by some but by no means all of the other donors. Until the mid-1990s Japan and France were the next largest donors (along with Germany) but their aid volumes have plateaued or dipped slightly and their relative ranking as major donors has fallen. Two of the countries to have risen in this regard are Germany and the UK, the latter's increases in ODA since 2000 being substantial.[3] Although not included in this graph, we can point to examples of two other donors to illustrate some interesting changes over time. South Korea has emerged from

Figure 3.4 ***Donors: ODA as a % of GNI 2017***

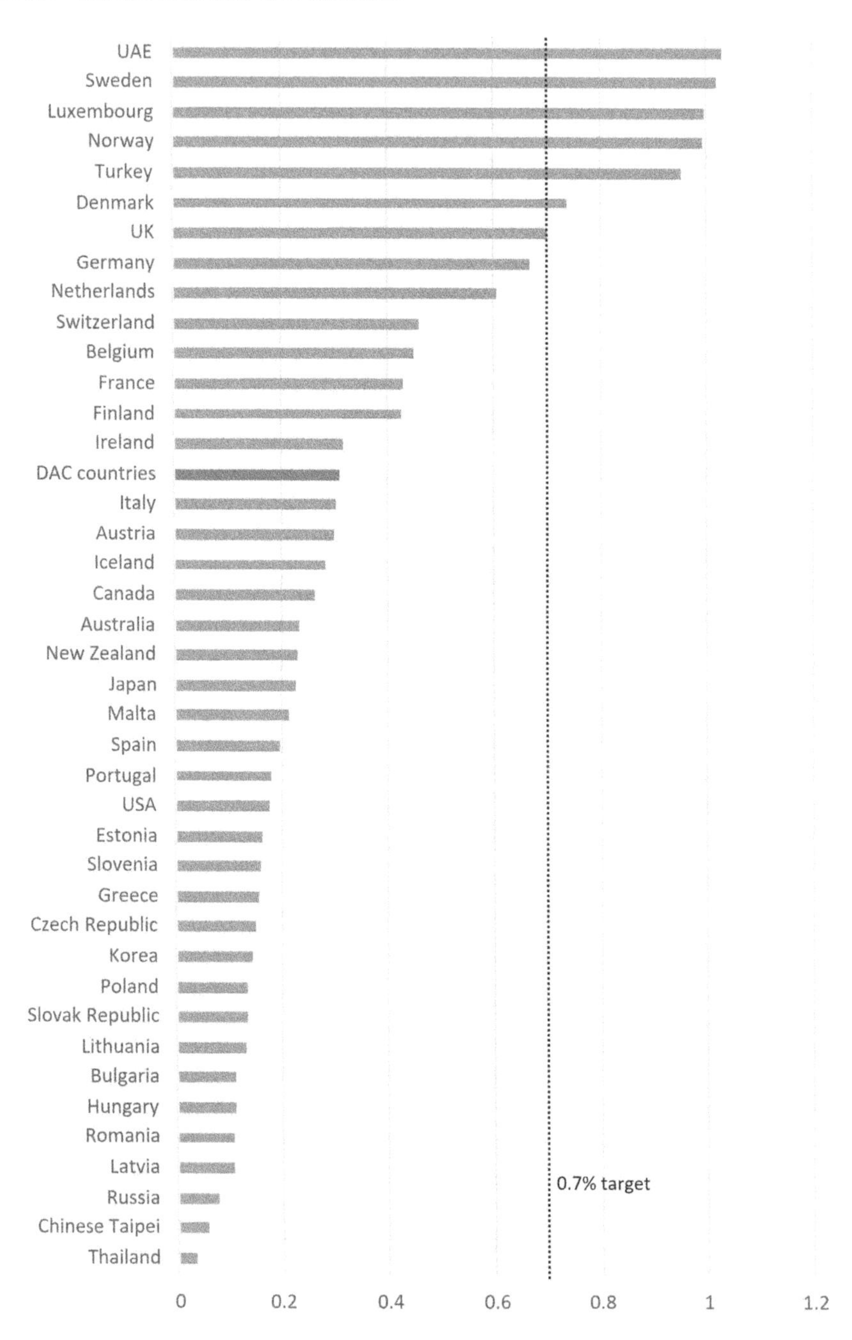

Source: https://data.oecd.org/oda/net-oda.htm

Figure 3.5 ***ODA from main DAC donors 1970–2017 (constant $US mill 2017)***

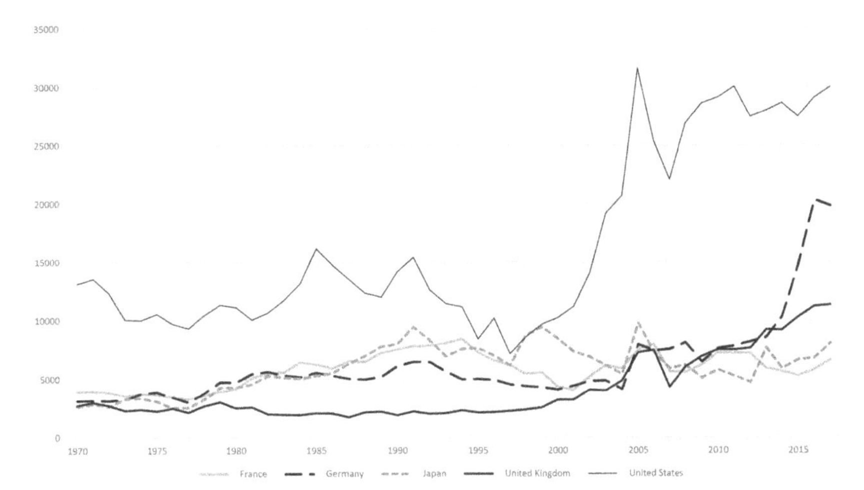

Source: www.stats.oecd.org

being considered a poor or developing country for much of the past 60 years to being a significant donor with rising ODA budgets since 2000. By contrast, Spain had followed an upward trend since starting ODA donations in about 1980 but the global financial crisis of 2007–08 hit it hard and its ODA contributions have fallen drastically since. In addition to the countries depicted in Figure 3.4, we should note that the list of DAC donors has expanded (see Table 2.2) with ten new members since 1990 but, as yet none of these has emerged to match the five largest established donors.

Non-traditional donors

Alongside these established and emerging donors within the OECD system, we can begin to gain an impression of how other countries are playing a role as aid donors and how rapidly this has been changing (Woods 2008; Six 2009; Greenhill *et al.* 2013; Eyben and Savage 2013). Because we are just analysing ODA at this stage, it is important to note that this gives us only a partial impression of these new players. Some are not included at all (most notably China) and we know that their development aid and co-operation activities are very substantial globally. Secondly, others appear only to have partial reporting of their aid budgets. Taiwan/Chinese Taipei does appear (see Figure 3.7), but the amounts seem very small compared to what we suspect anecdotally to be significant forms of assistance to countries with which it has diplomatic relations.

Figure 3.6 ***Main non-DAC donors: average annual ODA 2013–17 ($US mill constant 2017)***

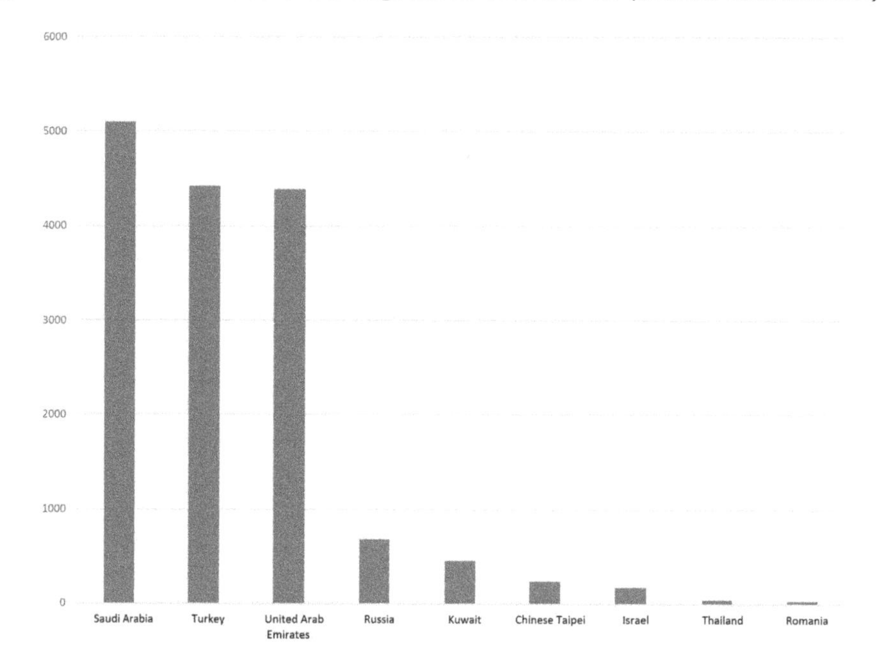

Source: www.stats.oecd.org

Figure 3.7 ***Main non-DAC donors: ODA 1970–2017 ($US mill constant 2017)***

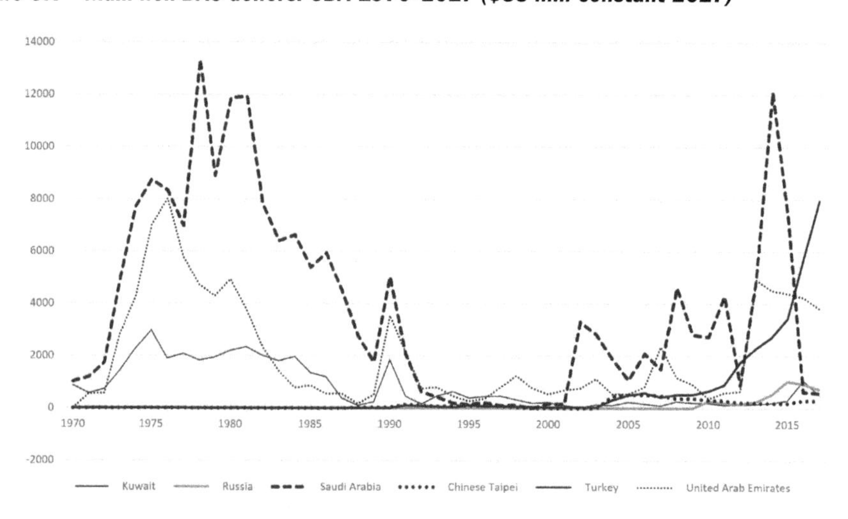

Source: www.stats.oecd.org

Nonetheless, the DAC system does reveal some interesting points about these non-traditional donors. Several oil-producing countries (Saudi Arabia, UAE and Kuwait) (see Figure 3.9) are revealed as important donors, Saudi Arabia ranking alongside Japan and France in terms of total volumes of ODA in recent years. There are also emerging donors, perhaps those who are eyeing membership of the OECD in future (Russia, Romania and Bulgaria) and who are beginning to comply with DAC's definition and data reporting procedures. Figure 3.6 shows the major non-DAC donors in terms of their recorded ODA in the past five years. Three countries dominate (Saudi Arabia, United Arab Emirates and Turkey), each with over $US1 billion of ODA each year, followed by Russia and Kuwait. It is probable that much of this aid is related to aid and refugee costs associated with the conflict in Syria. Yet beyond these five largest non-DAC donors, ODA volumes are very small: the other 15 donor countries recorded totalled an annual average of only $US 622 million in 2013–17.

Interestingly, these aid flows from non-DAC donors have some historical antecedents. Looking at flows since 1970 (Figure 3.7), we can see that the oil-producing countries of Saudi Arabia, Kuwait and United Arab Emirates were significant donors in the 1970s (probably a consequence of the very high oil prices in that decade). Indeed, in the second half of the 1970s, ODA from the non-DAC oil-producing countries amounted to as much as two-thirds of the total aid volumes of all DAC donors. Aid fell away in the 1980s and 1990s but Saudi Arabia and UAE re-emerged as large donors after 2000, joined by Turkey after 2010.

If we move outside the DAC ODA framework however, the picture becomes much more complex. China has become a significant aid donor, though perhaps we should note that it sees itself more as a 'development co-operation partner' than and 'aid donor'. It does offer grants for development purposes (see Box 3.1 for examples in the Pacific Islands) and it has offered a great many scholarships for students to study in China and develop essential skills as well as learn the Chinese language and customs. However, the bulk of its development assistance is in the form of concessionary loans. The precise amounts of these loans and terms remain largely hidden from the public eye and result from negotiations with separate recipient governments but we know that they are very substantial. They are mostly concerned with providing funds for construction and infrastructure (government buildings, sports arenas, roads, harbours, telecommunications etc.) and there is evidence of major Chinese-funded road projects throughout Africa, Latin America, Asia and Oceania. Another feature of these loans is that they are apparently forms of 'tied aid' in that they are linked to agreements for Chinese firms to be directly involved in the provision of facilities. Thus, Chinese loans are provided (at seemingly favourable rates) so that recipient governments can have roads and buildings constructed for them by Chinese companies (using Chinese equipment and employing mostly Chinese labour). Although this falls outside the purview of DAC and would seem to be an overt form of tied aid with clear benefits back to the donor, they are now a very important feature of aid, globally. They are clearly a form of 'co-operation' which makes explicit that there will be benefits for both sides (Sears 2019). And because the Chinese government is ready to fund such projects on a large scale, it is proving to be a popular option for many recipient/partner governments who find the Chinese donors more straightforward to deal with than their less generous Western counterparts. What is not yet apparent is how the loan repayment process will proceed. The loans are provided through Chinese (state-owned) banks and it is clearly expected that they will be repaid at the agreed rates – they are commercial agreements, albeit at below market rates. However, it is also likely that there will be some instances when problems with repayment will be met with a favourable response by the Government of China, either forgiving some debt or allowing some rescheduling. In other instances, though, repayments of mounting debt burdens by some countries may become problematic and may lead to difficulties in the future.

Box 3.1 China and the Pacific Islands

Although the Pacific Islands receives a very small share of development assistance on a global scale, and from China (PRC) there are much smaller flows compared to Southeast Asia or Africa, the region and China's involvement in it has received much recent attention. USA, Australia and some European donors have taken a renewed interest in Pacific Island countries after seeing major Chinese-funded infrastructural projects in the region. Some commentators have raised concerns about the level of indebtedness to China and the rising geopolitical influence it has in the region (Brant 2016; Hanson 2008, 2009).

As well as the conspicuous infrastructural projects (roads, ports, telecommunications, energy, water reticulation), there have been projects to provide new government buildings and sports facilities, all usually involving low-interest loans from China and the use of Chinese companies to build the facilities. There have also been grants, often in support of the larger loan-funded projects, and there has been a big increase in scholarships for Pacific students to study in China.

Yet it is very difficult to quantify these flows and China depicts them, with justification, not as aid projects but as forms of development co-operation as it seeks to extend its economic and political influence in the region, and many Pacific Island countries have been happy to tap into a new source of development financing. An exercise by the Lowy Institute in Australia attempted to gather a range of data on the projects, largely from local media reports in the region. Its database, although not complete and not including a range of non-material and other assistance such as scholarships, does provide a good insight into the nature and volume of Chinese 'aid' in the region.

Firstly, it is apparent that if we just look at grants – direct funding for aid projects not involving loans – that China is a relatively small donor in the region. Using Lowy Institute data for 2010–14, Overton *et al.* (2019: 118) showed that China disbursed the equivalent of about $US 35 million per year in the region. That compares with the largest DAC donors of Australia ($US 916 million), New Zealand ($US 136 million) and Japan ($US 134 million).

However, China tends to use concessionary loans rather than grants and this reveals an interesting shift. In the region, the other main providers of concessionary development loans are the World Bank and the Asian Development Bank (ADB) and they fund similar infrastructure projects. Again the comparison is not exact, for we do not know the detailed terms of many of these loans, but we can compare figures for loans from China against those from the World Bank and ADB (Figure 3.8).

Figure 3.8 ***Concessionary loans from China and development banks to Pacific Island countries 2010–14 (annual average $US mill current prices 2014)***

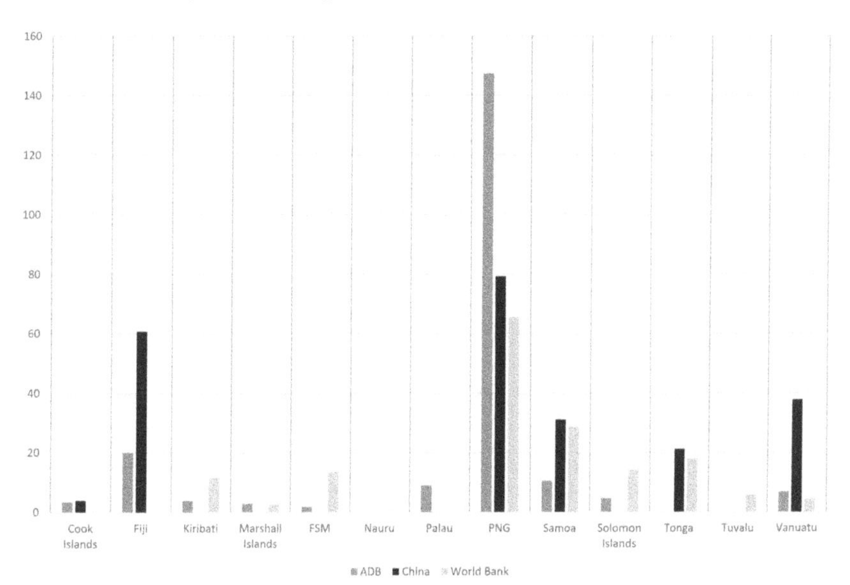

Source: Overton *et al.* 2019: 119 from OECD.Stat (www.stats.oecd.org) and Lowy Institute data (https://chineseaidmap.lowyinstitute.org/)

Overton *et al.* (2019: 120) noted:

> *[China's] average of $US 234 million exceeds both the ADB ($US 210 million) and the World Bank ($US 165 million). Furthermore, given the smaller number of countries it lends to, it has become by far the largest lender to Fiji and matches the World Bank in Tonga and Samoa and the ADB in Cook Islands. Yet … it is apparently only present as a lender in five of the twelve Pacific countries recorded as concessional loan recipients.*

Figure 3.9 ***Chinese aid in Tonga: construction of new government buildings***

Photo: John Overton

Furthermore:

> *China has become a major actor in the Pacific aid landscape. It is far from a traditional 'aid' donor, though its grants are locally important, if relatively minor in aggregate. Instead it operates more as a large commercial assemblage: its government provides very large loans to certain countries at concessional rates, largely for big infrastructure projects, and these are used in large part to pay Chinese corporations to undertake construction. Rather than being greeted with alarm and cynicism, perhaps such operations should be seen (with caution) as a form of development co-operation that provides the region with resources for desired projects that are not always available from traditional donors, especially not through the bilateral programmes of established donors … In this sense, we*

should see China as sitting alongside the ADB or the World Bank as a significant lender, rather than framing it as an 'aid donor'. As such, China may well impose its own (possibly opaque) conditionalities in terms of economic or political concessions but, we might suggest, this may not be that different (in scale if not nature) compared to the loan conditionalities historically imposed by the IMF, World Bank or ADB.

Other non-traditional donors are likely to emerge in the future. Of interest here are those countries in transition, formerly or still aid recipients but with rapidly growing economies and ambitions and plans to extend their diplomatic and economic interests worldwide. Thus India, Brazil and South Africa have signalled that they will expand their aid operations in future. India is clearly reinforcing its own commercial and diplomatic interests through a strategic use of aid (Fuchs and Vadlamannati 2013) and Brazil has strengthened its relationships in Africa through the use of technical assistance (Abdenur 2015). We can expect these new donors to operate along similar lines to China, seeing aid as an element of South–South co-operation and a way to seed and support their trading and investment interests internationally.

We will see later in this book that these sort of trends in the donor world are likely to operate more and more outside of the OECD 'club' and conform to more recent concepts 'shared prosperity' as new donors seek their own ways of operating, more explicitly pursuing their self-interest and less willing to conform to global agreements and harmonised ways of working.

Multilateral agencies

Multilateral aid agencies consist of a wide variety of international agencies. Firstly, we note the European Union. Although European countries have substantial bilateral programmes, they also pool ODA resources to operate collectively. This pool is larger in size than any single member's contributions. Then there is the United Nations group of agencies, itself funded by UN members to varying degrees. These agencies include bodies directly concerned with aid and development (such as United Nations Development Program

(UNDP)) or relief efforts (United Nations High Commission for Refugees (UNHCR)) and many whose specific roles touch on particular development issues (United Nations Children's Fund (UNICEF), Food and Agriculture Organisation (FAO), World Food Program (WFP), United Nations Environment Program (UNEP), United Nations Programme on HIV/AIDS (UNAIDS) etc.).

There is then the group of development banks, headed by the World Bank group and including a variety of regional development banks (ADB, African Development Bank, Inter-American Development Bank etc.). These banks, together with the International Monetary Fund (IMF), grew out of the Bretton Woods Agreement of 1944 and had the mission of supporting reconstruction and development in the global economy after the Second World War. Other development banks have emerged separate from the Bretton Woods framework. These include the Arab Bank for Economic Development in Africa, BancoSur (South America), the Council of Europe Development Bank and the Islamic Development Bank. The resources provided by these agencies are counted as ODA when they meet the criteria for aid, though this is less than their total operating budgets. Once again, we need to note that, like much of the assistance provided by the PRC, these development banks offer some direct grants but most of their activities involve concessionary loans, in this case where the credit-worthiness of 'donors' is used to secure favourable interest rates for borrowing countries that would otherwise face higher market rates because of their higher risk profiles.

To these multilateral institutions have been added a range of agencies which have a particular development role, such as the Global Alliance for Vaccines and Immunization (Radelet and Levine 2008) and the International Atomic Energy Agency. In recent years there has been a growth of multilateral funding agencies focused on different aspects of environmental change. These include the Adaptation Fund, the GEF, and the Climate Investment Fund. Together these have become significant features of the aid landscape in the past decade.

Private and corporate donors

Finally, we see those aid donors which mainly draw on private and philanthropic funds. It is a diverse array that ranges from Coca-Cola® to Comic Relief. Some operate as multilateral agencies. For example, the Global Fund brings together a diverse range of funding sources, including governments, private philanthropists, private sector companies and NGOs (Radelet and Levine 2008). Raising nearly $US 4 billion a year the Fund focuses on the elimination of key diseases: AIDS, tuberculosis and malaria. There are the large private philanthropic donors, such as the Bill and Melinda Gates Foundation (Fejerskov 2015), which follow the philanthropic traditions and models of the Ford Foundation and the like. There are also many agencies which provide 'corporate aid'. These are private sector businesses which donate finds for development, often under the principle of corporate social responsibility. Some, such as Nestlé, have distributed aid, often in ways which link the geography of their international commercial presence with patterns of their aid distribution (Metzger *et al.* 2010). And, of course, there are development NGOs: a huge array spanning faith-based organisations (World Vision, TearFund, Caritas etc.) and more secular ones (Oxfam, CARE, MSF, etc.).

However, whilst such organisations are often visible in donor public consciousness in terms of aid and relief, only a few (such as the Bill and Melinda Gates Foundation) are seen in the statistics for ODA. Private donors have been, and continue to be, an important element of the aid world. Many are associated with long-standing and well-grounded networks and development activities (Wallace *et al.* 2007). Yet, compared to the budgets of the larger bilateral and multilateral agencies, only a handful can be considered to be significant donors at the global scale.

Summary of ODA donors

The operations of many multilateral agencies (though not so much of the more private philanthropic agencies or development NGOs) are able to be analysed through the DAC system and putting together these data with the bilateral donors we looked at above, we can obtain an interesting picture of the overall sources of significant ODA (Figure 3.10). Here we show the major ODA donors, those contributing an average of over $US 200 million per year over a recent five-year period (2013–17).

Some key features emerge from these data. Firstly, we note that the large DAC donors – the big five (US, Germany, UK, Japan

Figure 3.10 ***Ranking of major ODA donors 2013–17 (annual ODA 2013–17 average $US mill constant 2017)***

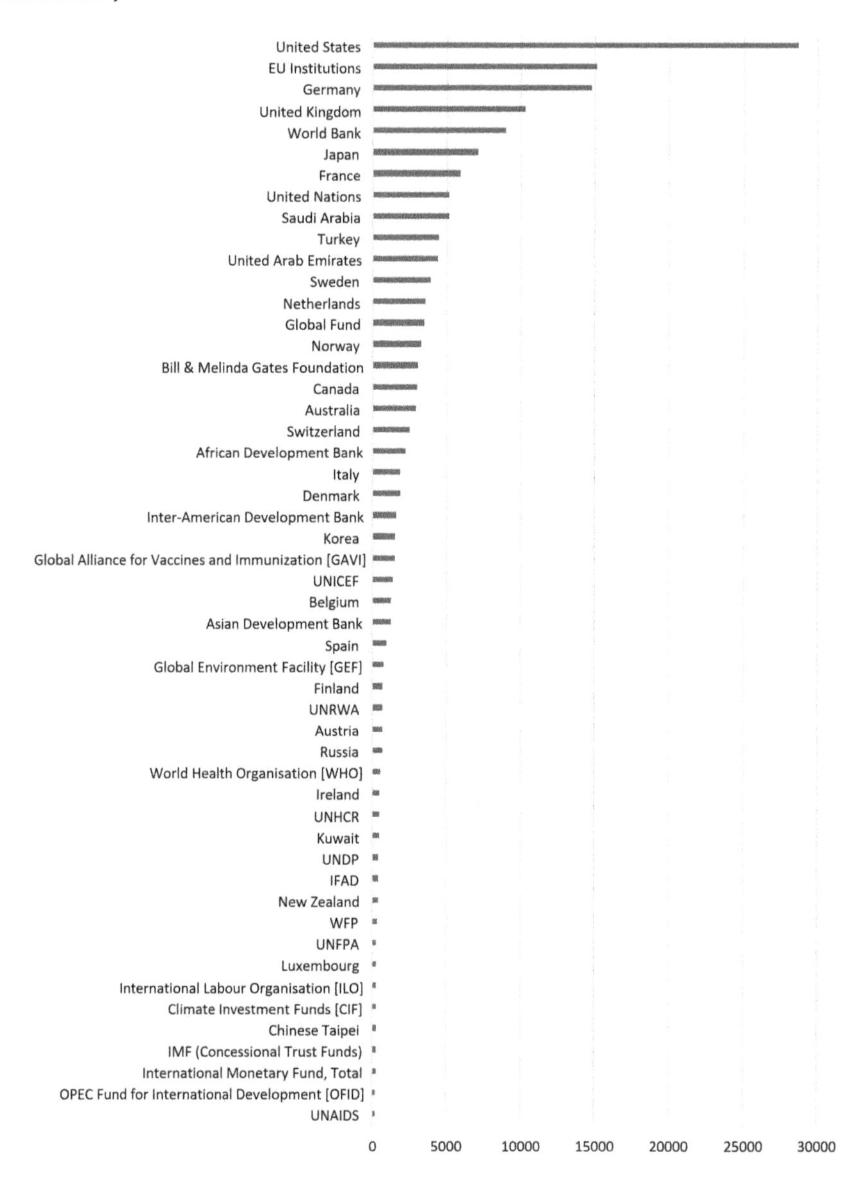

Source: www.stats.oecd.org

and France) – are still prominent, accounting for five of the seven top-ranked donors. Below them OECD donors (mostly members of DAC) are well represented. These donors are interspersed with the range of multilateral agencies, most notably the EU, World Bank and UN in the upper ranks, and the Global Fund, the Gates Foundation and the various development banks down through the order. However, perhaps the most interesting feature of this diagram is the prominence of a number of non-traditional donors. Their ODA efforts have been counted by the DAC even though their activities lie outside the DAC framework and agreements. The oil-rich countries of Saudi Arabia, especially, UAE and Kuwait are the most notable but we also see important contributions from Russia and Taiwan (Chinese Taipei) (bearing in mind that their total aid efforts are probably not well accounted for through this ODA analysis).

Types of aid

Whilst donors give varying amounts over time and there are marked differences in the volumes of ODA they disburse, there are also significant differences in the types of aid that different countries and institutions support. We can see a broad range of sectors that ODA goes to. There is a broad welfare category, which includes many of the basic sectors that many people associate with aid: education health, population and reproductive health and water and sanitation. There is a category that has to do with building public institutions and their capacity (government and civil society) and building peace and security. There is then economic infrastructure (transport, communications and energy) and the 'productive sectors' (increasing production and productivity in agriculture, fisheries, forestry, mining, tourism, construction and industry, and developing economic and trade policies). Multi-sector and 'cross-cutting' activities include the promotion of environmental protection and gender issues whilst general budget support and debt action usually involve direct support for governments either supporting the everyday costs of state activities or retiring debt. Humanitarian aid and food aid/security are usually related to particular emergencies or issues of shortage. Finally, there is often a large 'unallocated' category of ODA. This is for expenditures which cannot be readily allocated to a sector but also includes support for development NGOs and administrative costs.

Taking a broad overview of these sector allocations (Figure 3.11) it is apparent that ODA is spread quite evenly over these many sectors. Education, health and other welfare activities account for less than a quarter of ODA (a surprisingly low proportion perhaps given their prominence in the MDGs and SDGs). 'Government and civil society' support is another significant category (along with humanitarian assistance and the 'other/unallocated' category) whilst economic

Figure 3.11 ***Sector allocations of ODA 2013–17***

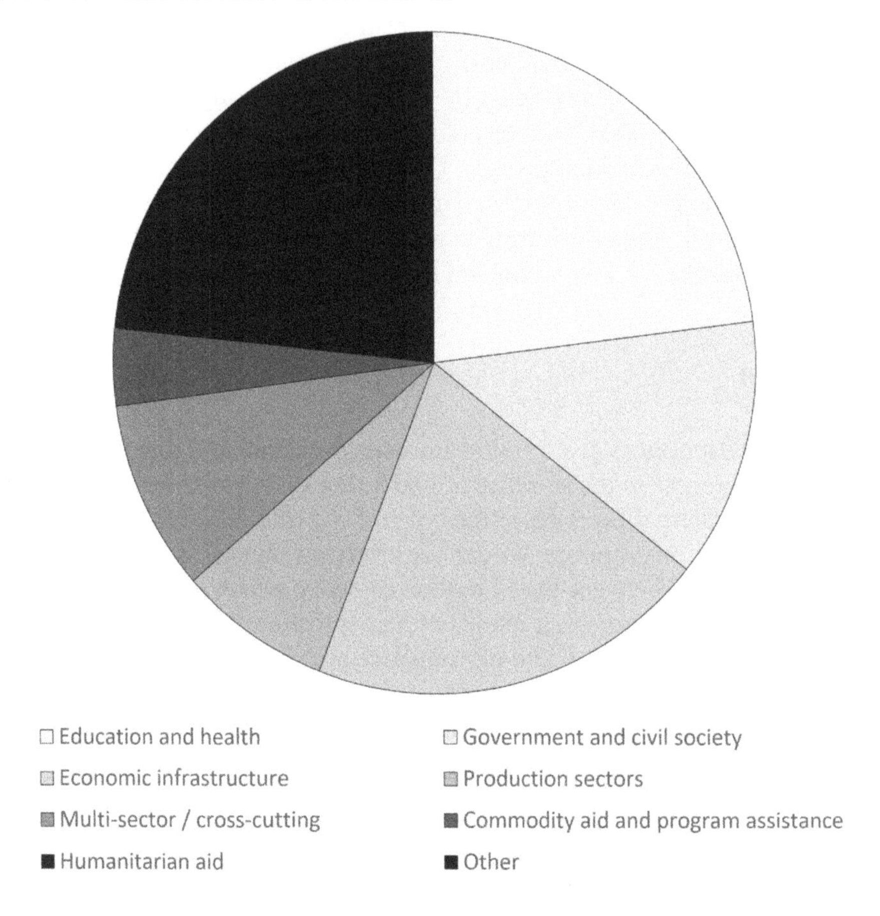

Source: www.stats.oecd.org

sectors (infrastructure and industries) together make up just over a quarter of total ODA. It is a diverse portfolio of sectors supported globally by ODA.

Adding to this overall diversity we can see significant differences in the way different donors have different sectoral priorities for funding (Figure 3.12). If we look at some examples of donors, including the main bilateral and multilateral agencies and two additional donors (Sweden and the Global Fund to illustrate some peculiarities of donors), it is apparent that donors develop their own strategies and decide to fund particular sectors rather than others.

Of the donors shown here, all support welfare provision in some way, though Sweden and Japan are relatively lower funders in this

Figure 3.12 ***Sector allocations of ODA 2013–17 by selected donors***

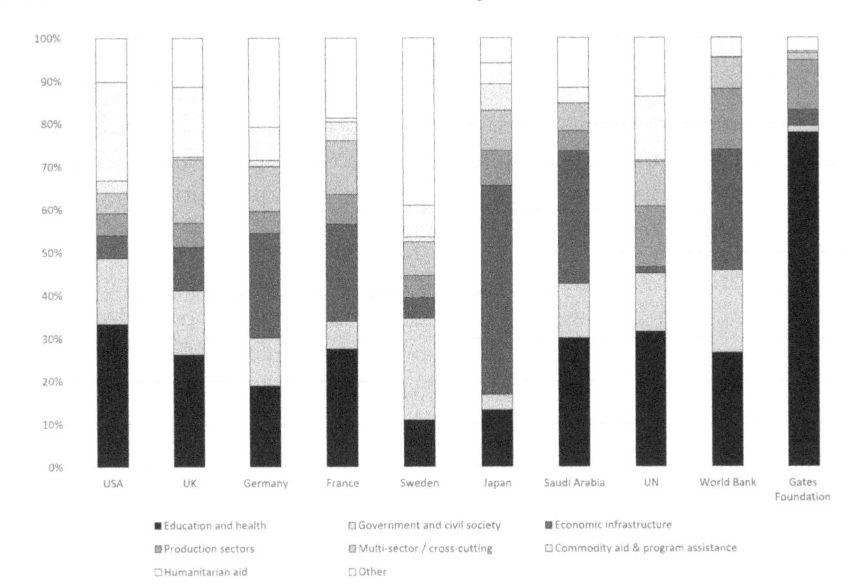

Source: www.stats.oecd.org
Note: Saudi Arabia data for 2015–17 only.

regard, whilst the Bill and Melinda Gates Foundation concentrates heavily on health-related activities. Support for building the capacity of government and public institutions seems to be of significant interest for most European donors, USA, the UN and the World Bank. On the other hand, economic infrastructure seems to be a speciality of Japan, although the World Bank also puts emphasis here, along with Saudi Arabia, France and Germany. Similarly, building productive sectors seems to be a particular concern for the large multilateral agencies. By contrast, the cross-cutting issues of the environment and gender are well supported by European donors and the UN but less so by USA. Of the remaining sectors, it seems that humanitarian assistance is supported by nearly all donors to varying degrees (the high relative and absolute expenditure by USA is notable here), whilst the relatively small amounts to programme support come from Europe. Unusually, Sweden has a very high unallocated amount, perhaps reflecting its strong support for development NGOs working across a range of activities.

We will return to the issue of contrasting donor priorities across different sectors in later chapters, and changes over time, when we examine changing theories, regimes, and practices of aid.

Recipients

Having mapped where ODA comes from and what it is spent on in terms of donors, we can now investigate where aid funding goes in terms of recipients. Here we see the dominance of country-based systems: countries, usually governments, receive and disperse aid funding. But there are also important flows to multinational (or regional) institutions as recipients.

Firstly, on a broad scale, it is apparent that the largest share of ODA goes to Africa, south of the Sahara – some 42 per cent globally – and this is the region where development indices show the most pressing need for assistance (Figure 3.13).[4] Interestingly, South and Central

Figure 3.13 ***ODA recipients by region 2013–17***

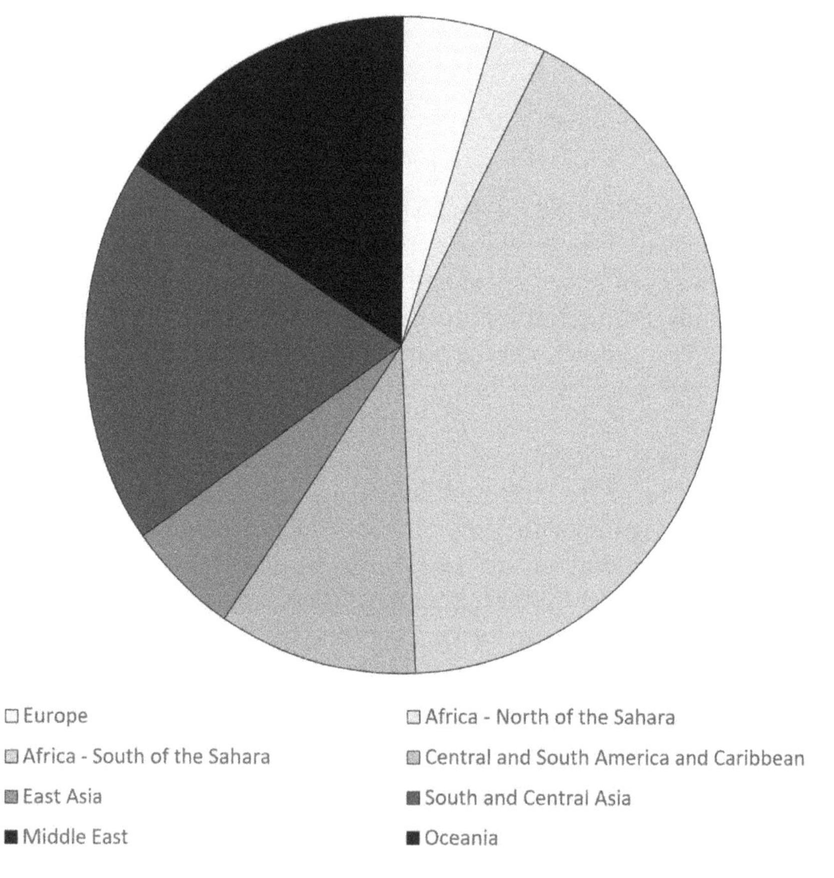

Source: www.stats.oecd.org

Asia comprise the next largest flow (with just over 20 per cent). This region includes one of the largest developing countries (India) and the largest single recipient of ODA (Afghanistan). This latter example, involving the way ODA often follows military intervention into areas of conflict, also helps explain why the Middle East (including Iran and Syria) is the next largest region.

These features can be seen more clearly when we look at the ranked order of recipients for the 2013–17 period (Table 3.1). This list

Table 3.1 ***Ranking of major ODA recipients 2013–17 (ODA disbursements average per year at constant $US mill 2017)***

Rank	*Recipient*	*$US mill*
1	Afghanistan	3519
2	Syria	2009
3	Vietnam	1961
4	India	1951
5	Ethiopia	1935
6	Pakistan	1588
7	Kenya	1576
8	Iraq	1556
9	Tanzania	1506
10	Jordan	1505
11	Myanmar	1460
12	Bangladesh	1446
13	South Sudan	1388
14	West Bank and Gaza	1261
15	Mozambique	1242
16	Nigeria	1233
17	Democratic Republic of the Congo	1154
18	Uganda	1044
19	Colombia	969
20	Morocco	906
21	South Africa	892
22	Somalia	804
23	Ukraine	786
24	Yemen	739
25	Turkey	699

Source: www.stats.oecd.org

includes all ODA recipients receiving an average of over $US 500 million per year. African countries dominate the list but there are notable inclusions of countries embroiled in conflict situations (Afghanistan receives twice as much ODA as any other recipient, also Iraq, Palestine (West Bank and Gaza), Syria and Ukraine are significant recipients) or deemed to have high strategic value (Pakistan and Jordan). Interestingly on this list there are also countries which have experienced fairly rapid economic growth and have quite reasonable development indices (Vietnam, Brazil and Mexico), but whose continued economic development seems to be well supported by donors.

We can suggest then that there is no clear correlation between how needy a country is and how much aid it receives. This may be the case for the presence of so many countries from south of the Sahara but it certainly does not explain the volume of ODA going to many other countries. A basic analysis of ODA measured against a country's development status (its HDI (Human Development Index) score – see Figure 3.14) shows this lack of a clear picture. Perhaps the

Figure 3.14 ***ODA and HDI 2017***

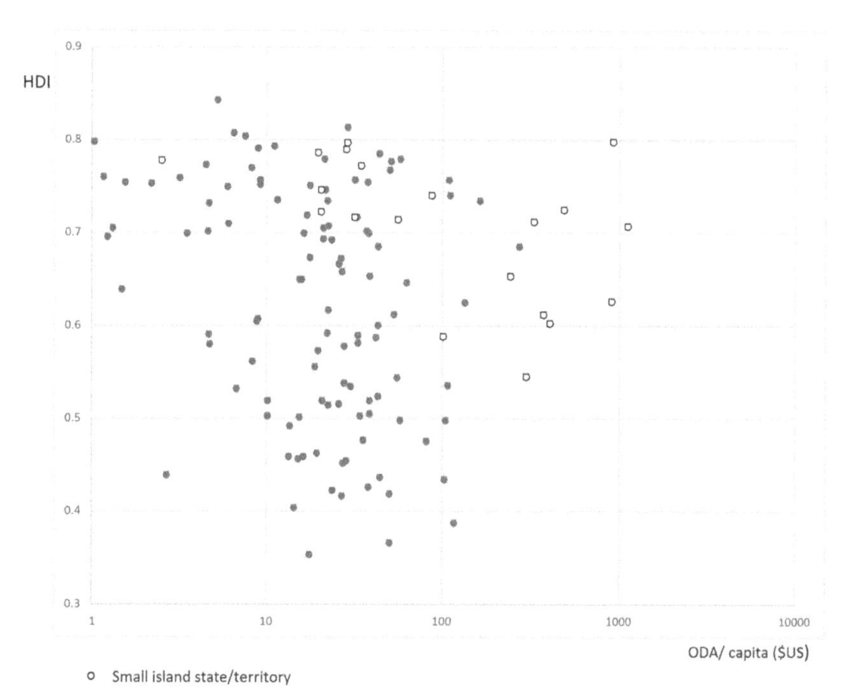

Source: www.stats.oecd.org

only feature of these data is that small island states and territories generally seem to be amongst the highest recipients of ODA per capita (as we have seen) but they also rank quite highly in terms of HDI.

Another way to see where aid goes is to look at how much aid countries receive relative to their population – and here the picture is quite different (see Table 3.2). The most notable feature of this table is

Table 3.2 ***Ranking of major ODA per capita recipients 2013–17 (ODA disbursements average per year at constant $US mill 2017, using UN estimates of 2015 population)***

Rank	*Recipient*	*$US mill*
1	St Helena	22612
2	Tokelau	14126
3	Wallis and Futuna	7649
4	Niue	7519
5	Montserrat	7496
6	Nauru	2029
7	Tuvalu	1907
8	Marshall Islands	1093
9	Cook Islands	1028
10	Palau	911
11	Federated States of Micronesia	900
12	Tonga	479
13	Vanuatu	402
14	Kiribati	369
15	Samoa	324
16	Solomon Islands	295
17	West Bank and Gaza	270
18	Cabo Verde	240
19	Jordan	164
20	Timor Leste	134
21	South Sudan	117
22	Mongolia	110
23	Lebanon	109
24	Syria	107
25	Afghanistan	104

Source: www.stats.oecd.org

that the top 16 and most of the world's top 25 aid recipients on a per capita basis are small island states and territories. We will examine this category of aid recipients in detail below.

Aside from the small island states, however, it is interesting to note that our high ODA per capita list includes a number of countries that also appear on the total ODA list (Table 3.1): Afghanistan, West Bank and Gaza, Jordan and South Sudan. These are countries that have attracted attention not just because of the poverty experienced there, but because they have particular strategic value or are associated with conflict or military intervention.

Putting these different issues and pieces of evidence together, it is possible to suggest a categorisation of aid recipients. These are broad and sometimes overlapping groups but they do reveal some quite different motives and strategies for allocating aid – a theme we will take up later.

High poverty countries

Many people would reasonably assume that international development aid is given to tackle poverty and meet pressing development needs. This is certainly the assumption behind calls to increase aid volumes and be guided by, firstly the Millennium Development Goals (MDGs) of 2000–15 and now the Sustainable Development Goals (SDGs) of 2016–30.

Looking at the data on the largest aid recipients (Table 3.1) we certainly see a high frequency of African countries south of the Sahara and others with relatively poor development indices (such as Haiti and Nepal). This group in aggregate accounts for the largest number of aid recipients and also for the highest total volume of aid flows. All major donors are involved (along with China and other non-traditional donors such as India) and their projects and programmes cover the full range of sectors we noted above, from welfare to humanitarian relief and, in some cases, debt relief. Some of these countries have been able to put in place some reasonably efficient and accountable public institutions that can directly handle increasing aid volumes through mechanisms such as SWAps (Sector-Wide Approaches) and GBS (General Budget Support) (see Chapter 4); others still lack the capacity at state level to take the lead in development activities; whilst others unfortunately straddle the poverty and conflict categories (e.g. Mali and Sierra Leone) and suffer the joint afflictions of very high levels of poverty, and dysfunctional or contested states.

If we compare ODA levels with development indices (in this case a country's HDI score[5]) we can map this category of countries in Figure 3.9. These are the countries that lie in the lower part of the diagram (with HDI scores less than 0.55 and categorised by the UN as having 'low human development'). Only 6 of the 41 countries listed in this low HDI category lie outside Africa.

Countries in conflict with external intervention

Some of the countries with the highest levels of aid (in both absolute and per capita terms) historically and at present have been those who have suffered from major conflict and the involvement of external powers. This intervention has been in the form of direct military intervention to overthrow unwanted regimes (as in Iraq and Afghanistan following the 9/11 attacks in 2001) or to protect regimes under internal threat (Iraq's battle against ISIS (Islamic State)). Military intervention in the form of peacekeeping operations have also been in evidence in places such as Mali, South Sudan or Timor Leste.

In these cases, aid has followed military intervention, sometimes in rather controversial ways. Military intervention has occurred not because development is seen to be needed but because 'regime change' is a priority. But then, once political change has occurred, it is often realised that forms of development are needed to secure the legitimacy of new regimes and win the 'hearts and minds' of the general population in conflict zones. Thus, schools, health centres, roads, bridges, community facilities, microfinance schemes and so on, supplied by aid projects, become tangible expressions of hoped-for order and progress. Development assistance then in these situations is seen as part of a long-term solution and an exit strategy for the military. ODA can help build an environment that supports the legitimacy of occupying forces and/or the new regimes they install. This has been one of the key features of intervention in Afghanistan since 2001 when 'provincial reconstruction teams' established by the occupying forces have attempted to combine the efforts of military, NGO and local government agencies to provide forms of development, stability and, optimistically, prosperity for local communities.

Aid workers, wittingly or unwittingly, become agents of new regimes backed or installed by military action and this can be uncomfortable and even dangerous. Another aspect of this form of aid is that it may well involve military personnel and equipment in development activities. Rebuilding a school or a bridge by military forces as part of peacekeeping and peace building operations takes development assistance into a new realm and may involve an accounting exercise that sees some military expenditure counted as ODA. On the other hand, the experience of Afghanistan and Iran has seen a number of military leaders recognise that development, poverty alleviation and, crucially, community participation are crucial if civil order is to be established.

Countries experiencing internal conflict and natural disasters

Similar dynamics exist in countries that have suffered from internal conflict, but where external military intervention has been absent or limited. The recent conflicts in Syria have been enormously destructive and violent and very complex from a political perspective. Russia and USA have been involved in some military action but have largely stayed away from putting large numbers of their own military or aid personnel on the ground. Instead it has been their proxies in the country – the various factions they have backed – that it is hoped will help prevail and restore order. Syria is perhaps an extreme example – and the number of refugees flowing out of the country has been huge – but we see other internal conflicts and also natural disasters creating situations where both humanitarian and development assistance is called for and given. The earthquake in Haiti in 2010 for example, led to a very large-scale international response.

Development assistance in these circumstances, whether resulting from civil conflict or natural disaster, has usually combined forms of emergency help for those both *in situ* and those seeking to leave as either international refugees or internally displaced persons. There are specialist agencies which respond in such situations, such as Red Cross/Red Crescent or UNHCR, but donor governments have also developed the capacity to mobilise resources quickly and send needed resources.

However, ODA is important here for it is not just an emergency situation that demands attention but also crucially the more long-term needs of peoples and communities to rebuild their lives, businesses and communities. Here more conventional development strategies and agencies are employed. Typically, countries in this category may experience large influxes of aid in response to crisis (the Boxing Day Asian tsunami led to a marked spike in global ODA in 2005) but then a gradual tailing off of aid donations back to normal levels. For some countries where internal conflict and displacement is seemingly endemic and political solutions absent, however, ODA may keep flowing at high levels for many years. This is likely to be the situation in South Sudan, Palestine or Syria for example.

Again, we need to note that aid in these circumstances may have a key role to play but it is situated within a very difficult environment where public safety is uncertain, government agencies may not function effectively and civil society may be in a fragile state. Development and aid can help rebuild people's lives and livelihoods but where new conflicts and disasters are possible, aid work and development results can be problematic.

Countries of high strategic importance

ODA is used as a way to build diplomatic relationships as well as promote development and we can see this being reflected in the large flows of aid to countries whose development needs may be lower than others but who have importance because of their strategic location or diplomatic alignment. Jordan, for example, is a relatively prosperous developing country (its HDI puts it in the 'high human development' category of the UN), but it ranks very highly as an aid recipient (in the top 25 in both total and per capita terms – Tables 3.1 and 3.2). This is probably because its location, bordering Israel, Saudi Arabia and Iraq, means that it is seen as a critical country whose political stability and relatively liberal global outlook acts as buffer and a peaceful example in a region perceived as prone to extremism and conflict. Pakistan also is seen as critical for those donors anxious about the instability and extremism they see in neighbouring Iran and Afghanistan. Pakistan ranks (just) within the 'medium human development' category and could justify high volumes of aid on the basis of its development needs but its position as the world's fourth ranked aid recipient country is much more the result of its strategic importance to Western powers.

In many cases, aid in these situations directly helps support governments in providing important services for their citizens. These can include education and health services, infrastructure and energy as well as promoting economic growth in particular sectors. Typically, aid programmes to such countries tend to involve long-term commitments to development programmes and the use of recipient state systems. Aid which helps maintain a stable government that delivers effective services and an economy that provides jobs thus can act as a bulwark against political conflict. Furthermore governments that receive generous amounts of aid may well be more inclined to support the diplomatic and political strategies of particular donors. This facet of aid, seeing it as an important instrument of donor foreign policy, is a theme we will explore more in Chapter 4.

Emerging economies

Another important category of aid recipients is that which includes countries whose economies are growing relatively rapidly. Again, these are countries whose HDI rankings would seem to suggest that they are not amongst those most in need of aid. They have relatively prosperous and growing economies, they are usually tied to the global economies with large export-oriented industries and they usually have reasonably stable governments. They are large economies. In this category we see countries, such as Brazil, South Africa, India and, in the past, China. In addition, we would include the economies of Vietnam, Kenya, Bangladesh, Chile, Mexico and Colombia. Looking at the list of the top aid recipients (Table 3.1) we can readily see that a large proportion of global ODA flows to these emerging economies. Aid, then, seems to be a way of lubricating the machinery of the global economy, ensuring those economies that have good prospects for development are encouraged and supported to continue to grow.

Aid is given to these countries partly for geopolitical strategic purposes – it is good to keep emerging economies on-side and align their growing prosperity and global roles with those of donors. But also there are sound economic reasons. Growing economies and involvement in the global economy develops trading partners for the future and aid can play a role here, helping to forge business partnerships, providing technology and knowledge from donor sources that can be built on and used more in future, and influencing the direction of economic development so that it fits with donor strategies. Furthermore, providing development assistance in an environment where there is stability, growth and effective local institutions and agents promoting development is much more likely to succeed compared to situations of conflict, minimal economic activity and dysfunctional local institutions. Emerging economies are a good investment both politically and economically for donors.

Small island states

Small island states face particular development issues: their remoteness and relative small population bases mean that maintaining a full and viable system of government is very difficult (Tuvalu, for example, has a resident population of fewer than 10,000 people yet it is a fully independent country spanning a vast area of ocean and with a nearly full complement of government departments). Furthermore, many face particular environmental and economic problems, ranging from a lack of arable land to the serious threat of climate change and sea-level rise. Many have willingly maintained close diplomatic and constitutional ties with the former colonial powers and have been able to secure both high and continuing aid incomes and also access to metropolitan labour markets (Bertram 2004, 2006; Hintjens and Hodge 2012).

What is especially notable from the data in Table 3.2 is the high level of per capita aid in various Pacific states and territories. Eight of the top 10 per capita recipients and 14 (of the top 20 are in the Pacific Islands. This is partly a legacy of colonialism in that region involving principally Britain (and later its 'settled' colonies of New Zealand and Australia) and France. A number of the territories listed, such as Wallis and Futuna, Niue and Tokelau have some form of constitutional connection to former colonial powers (France and New Zealand in the case of the latter two). In the case of the Marshall Islands and Micronesia, the important strategic geopolitical location for the US in the Cold War and in the current political moment in the Pacific plays an central role (indeed were flows to American Samoa counted as aid this territory too would be highly ranked in these terms). Some of these territories such as the Cook Islands and Niue for example have relatively high levels of human development. Others such as the Solomon Islands and Kiribati do not however, and face serious developmental challenges associated with conflict and the environment respectively. Overall however, the high relative receipt in that region is probably more due to the combination of current and historical geopolitical patterns rather than the real challenges these societies face.

Overall, though, many small island states receive high levels of aid per capita. These states and territories, mostly lying in the Pacific, Caribbean and Indian Oceans, dominate the list of high per capita aid recipients (Table 3.2). Yet, perhaps even more than other recipients (for example, those in the 'emerging economies' or 'strategic importance' categories) these small island states do not rank poorly in the human development score. With the exception of Solomon Islands (one of the larger island states of the Pacific and with a recent history of internal conflict), and bearing in mind that many small states do not have a reliable enough statistical base to be able to calculate a meaningful HDI score, the small island states in Table 3.2 all rank in the medium or high human development categories. Although their export economies may be of questionable viability, the combination of aid and remittances has often provided jobs and incomes and led to the foundation of relatively good health and education facilities, improving literacy rates and life expectancy. Several now face the prospect of graduating to middle income status and are thus no longer eligible to receive assistance classified as ODA. For countries, such as Cook Islands, this may present a threat to a major source of funding for development and core government activities (Bertram 2018).

On the other hand, the case of St Helena and its inclusion in this list also rather stretches the credibility of the 'developing country' criteria and the definition of ODA (Box 3.2).

Box 3.2 St Helena: aid and airport development in an island territory

The island of St Helena lies in the South Atlantic Ocean. Without an indigenous population, it is a British overseas territory with a largely British-descended population with full British citizenship. Since British control was declared in 1657, it has long served as a strategic outpost and a link in global shipping and supply networks.

The island appears on the list of ODA recipients and, strikingly, it is recorded as receiving the most aid per capita (averaging over $US 28,000 a year for every one of its 4,500 or so residents – see Table 3.2). This level of aid alone, would see it classified as a 'high-income country', yet the funds spent on it by the UK government are still recorded as 'aid'.

The very large sums classified as aid were made up largely of nearly £300 million spent on building a new airport. 'The new airport was meant to improve accessibility and boost tourism, with the intention of making the island self-sufficient' (Dominiczak 2016) and therefore seemed to have some element of a development goal. However, given the island's past naval strategic importance, and the role played by the airfield at Ascension Island (1,300km to the north) in the re-taking of the Falkland Islands by Britain in 1982, no doubt potential military uses were also in mind.

However, the expensive new airfield has faced controversy. Problems with severe wind shear and fog has meant that commercial flights are marginal and only a limited service is in operation linking the island to South Africa. The influx of hoped-for tourists has not eventuated and the airport project received severe criticism from the British Parliament's Commons Public Accounts Committee in 2016 Dominiczak 2016).

We suggest that this is an example of rather dubious practices with regard to the inclusion of expenditure within the ODA framework that primarily relates to the maintenance of metropolitan strategic presence in remote island territories. It has little or no justification in terms of assistance for 'the economic development and welfare of developing countries as its main objective' (OECD 2019b).

Conclusion

Aid flows between donors and recipients in a great variety of ways. There exist relatively simple bilateral relationships between a donor and recipient governments but these, in aggregate, criss-cross the globe in highly complex ways. As such, many recipients will receive diverse forms of aid from a multitude of donors, each with their own economic, political and strategic priorities. Adding to this complexity is the large number of multilateral institutions and flows, many with specific mission goals and ways of operating.

Overall, there are certainly some dominant ODA flows: the major donors are prominent and far outweigh the large number of smaller and newer donors. Thus USA, Japan, and the larger European countries (together with China outside of the ODA framework) alongside the World Bank and the UN, are the major players in the donor world. Yet their activities are not evenly spread. The geography of aid is such that some donors, large or not, dominate in particular parts of the world and this is often related to a mixture of historical geopolitics and current strategic economic and geopolitical interests. Japan, not surprisingly, has a particular interest in Asia; Australia focuses on both Asia and Oceania; Canada has a strong interest in the Caribbean etc.

These strategic priorities of donors also mean that there is no simple correlation between recipient 'need' (as measured by HDI or GNI per capita) and ODA receipts. Many aid recipient countries receive large amounts of ODA because of their strategic or economic interest to donors, whilst some of the poorest countries languish with low per capita ODA receipts. The reasons why these inequalities exist has to do with the varying motives of donors and recipients, something we turn to in Chapter 4. However, firstly it is important to see how aid strategies have evolved over time, for not only is the geography of aid uneven and complex, its history has been marked by sharp changes of direction in terms of the rationale for aid, who receives it and how it is delivered.

Summary

- The geography of aid flows is highly uneven – there is no simple dualistic donor and recipient model and there is a complex range of historical, geopolitical and contemporary factors explaining this unevenness.
- The established donors operate within the DAC of the OECD. This group has grown over time as countries have moved into the ranks of 'high income' as defined by the World Bank.
- The DAC group, together with the established and linked agencies of the UN and the World Bank, continues to represent the major collection of donors and is a major force in terms of flows and agendas.
- Notwithstanding the continued role of the DAC there is much variability within its ranks in terms of which countries aid is given to and what sectors are targeted. Aid relationships almost always reflect historical ties and contemporary economic and strategic interest.
- There has been a rise in non-traditional donors. This has led to a widening of both donors and recipients as well as a widening in sectors targeted and the types of aid given.

- There is a core of very large donors that account for a significant proportion of all aid – this includes the USA, UK and Germany, among others. When we look at aid donors in terms of per capita outflows, however, the picture is different. The Scandinavian countries have been the most generous.
- There is no straightforward correlation between 'need' as defined by per capita income or the HDI and levels of per capita aid. There is a wide range of factors at play.
- We group recipients into six categories on the basis of their history and current nature – high poverty, internal conflict, external conflict, strategic importance, emerging economies, and small island states.
- Generally, lower middle income countries receive more per capita aid than the very poorest which suggests that economic rationale for aid flows is an important factor.

Discussion questions

- Which countries are the traditional aid donors and how has this changed over time?
- What are the problems with focusing on the DAC definition of aid in terms of understanding the motivations and impacts of aid relationships?
- Discuss the differences between traditional and non-traditional donors in terms of the way that aid is given and the things it targets.
- What are the historical and geopolitical factors that determine and aid relationship?
- To what extent does aid per capita correlate with 'need'? What factors complicate a straightforward correlation?
- What are the six groups of aid recipient types identified in this chapter? Describe each and give an example in each group.

Websites

- OECD (OECD.Stat) statistics on ODA: https://data.oecd.org/oda/net-oda.htm
- The Human Development Index – meaning and data sets: http://hdr.undp.org/en/content/human-development-index-hdi

Notes

1 Originally in the 1960s a target of 1 per cent of GNI in each member country was subscribed to but this included both government and private donations to developing countries. Given the difficulty of predicting or managing private flows, a focus just on ODA and a target of 0.7 per cent of ODA to GNP was adopted by the United Nations in 1970. It has been much referred to but rarely achieved by individual countries, and overall, since.
2 This high European ranking for ODA does not take into account the substantial amounts of ODA given by European Union institutions. Thus the 'generosity' of Europe is even greater than Figure 3.4 indicates.
3 Germany's dramatic rise in 2015 is due to the inclusion of refugee-related expenses. Note also that the spikes for many donors in 2005 was probably related to humanitarian relief for the Asian tsunami.
4 In this analysis we have not included a large amount of ODA, which is classified in regional terms as 'unspecified'. This goes to international institutions and activities where it is difficult to track exactly where in the world the funds end up. It also includes refugee and other costs which are spent within donor countries.
5 The Human Development Index (HDI) is a statistical measure that combines life expectancy, education (years of schooling) and per capita income (GNI at purchasing power parity) to gain a composite indication of a country's development. The Index varies between 0 (no development) and 1 (very high development) and is used to rank and track countries over time.

Further reading

Alesina, A. and Dollar, D. (2002) 'Who gives foreign aid to whom and why?', *Journal of Economic Growth* 5, 33–63.

de Haan, A. (2009) *How the Aid Industry Works: An Introduction to International Development*. Kumarian Press, Sterling.

Gulrajani, N. and Swiss, L. (2019) 'Donor proliferation to what ends? New donor countries and the search for legitimacy', *Canadian Journal of Development Studies / Revue Canadienne d'études du développement* 40(3), 348–368.

Mawdsley, E. (2014) 'Human rights and South–South development cooperation: Reflections on the "rising powers" as international development actors', *Human Rights Quarterly* 36(3), 630–652.

Overton, J. Murray W. E. and McGregor A. (2013) 'Geographies of aid: a critical research agenda', *Geography Compass* 7(2), 116–127.

4 Trends in aid

Learning objectives

This chapter will help readers to:

- Understand the history of contemporary aid patterns and colonial antecedents
- Define the concept of an aid regime and its component modalities
- Distinguish between the various aid regimes that have characterised the aid sector since 1945
- Appreciate the background, theoretical underpinnings, key agencies, policies and practices, modalities, and effects of successive aid regimes
- Critically appreciate the overlaps and interactions within and between the aid regimes discussed – modernisation, neoliberalism, neostructuralism and retroliberalism

Introduction

Having examined the way aid flows over space, with flows between the myriad of donor agencies and recipient states and organisations, we now turn to the ways these flows have changed over time. Although we might see aid as a relatively straightforward concept, involving the assistance offered to promote the economic growth and welfare of people in the developing world, in fact, ideas regarding why aid should be given, where it should go, how it should be delivered, and in what form, have changed markedly over time. This has resulted from changing conditions in the global economy, from changing global geopolitics and also from changing concepts and theories about what development should be and what role aid can play in this.

In this chapter we trace changes in aid thinking and practice over time and one way we can summarise and categorise these changes; through the concept of 'aid regimes'. We concentrate on the period after 1945 and on postcolonial settings, though we begin by looking at some colonial antecedents for models of aid and the consequences of these.

Aid over time: changing trends and common concerns

Although in this chapter we emphasise change and the evolution of different ideas, we should note at the outset that there are also some important consistencies over time. Firstly, it is notable that aid has been a remarkably constant feature of the global economy and global politics for the past 60 or more years. Rarely has it been suggested that aid should end altogether or be reduced substantially so that poorer countries should just fend for themselves. Aid seems to be a relatively permanent fixture in the theory and practice of global geopolitics. Secondly, a consistent, if not always transparent, thread has been that aid brings, and should bring, some benefits to donors as well as recipients. This view sits uncomfortably with many who support the altruistic concept of aid but the political realities are such that donor countries cannot readily sustain a large aid programme simply by telling their constituents that they are 'doing good' for those overseas, when there are many pressing needs for state funding at home. Perhaps of greater significance is the fact that political leaders in the Global North see that aid budgets can be used creatively to support wider strategic, diplomatic and economic goals of the donor country itself. As such, aid has become an integral instrument of foreign affairs and economic development rather than simply a separate and solely philanthropic activity. Aid is, and nearly always has been, about self-interest as much, if not more than it has been about altruism. Finally, a consistent element of aid is that its allocation is not determined by any sort of dispassionate assessment of need – it does not automatically go where it is needed most. Rather, as was discussed briefly in earlier chapters, its allocation is determined by a complex web of historical, political, diplomatic and strategic considerations that are rarely transparent.

To set the scene for this discussion, we can see how aid volumes have changed over the past 50 years (Figure 4.1). This recaps and summarises the discussion we undertook in Chapters 1 to 3. The figure shows flows of Official Development Assistance (ODA) in

Figure 4.1 ***ODA disbursements 1966–2017 ($US mill constant $ 2017)***

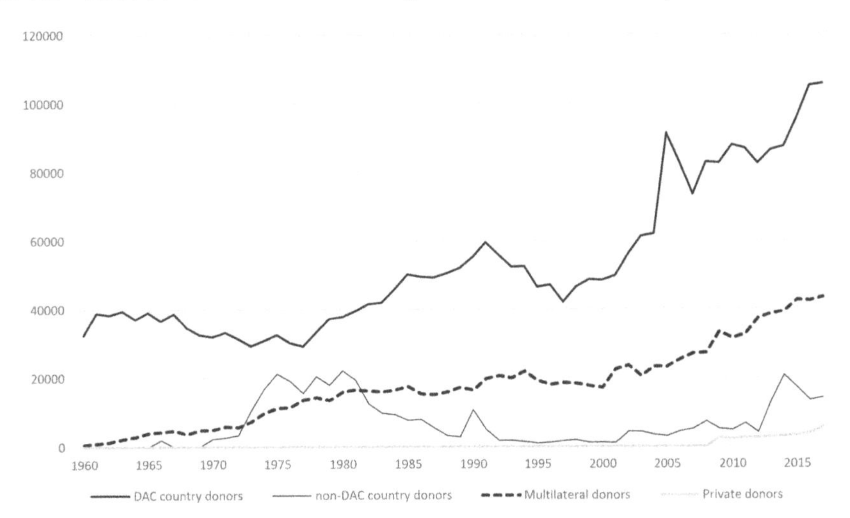

Source: www.stats.oecd.org

constant 'real dollar' amounts (allowing for inflation) using data defined and collected by the Organisation for Economic Cooperation and Development (OECD), though we should bear in mind the changes in the way ODA has been measured as well as the range of donors and recipients over this time period (Chapter 1). Moving through the periods, there seems to have been a gradual fall-off in ODA from the Development Assistance Committee (DAC) donors from the later 1960s to the late 1970s but a big increase from non-DAC donors from the early 1970s. These non-DAC donors at the time were mainly Saudi Arabia and Kuwait, both with large surpluses to expend following the dramatic increases in the price of oil during the decade. Throughout the 1980s ODA increased from the DAC donors (with declines from non-DAC donors), but then there was a reversal in the first half of the 1990s followed by some stabilisation and slight increase until 2000. Thereafter, a notable and steady increase took place until 2011, with a spike in 2005 coinciding with the Southeast Asian tsunami. In the last four years prior to 2017 the increases have continued and we have seen the appearance of new donors (the 'others' here being new private donors, principally the Gates Foundation). Overall, the DAC donors and the large multilateral organisations dominate ODA and only recently has the non-DAC group begun to increase its relative share (apart from the noted big contributions in the 1970s).

Figure 4.1 gives us the basis for the framework of aid 'regimes' we explore below – though as we will see common themes, together with the different unfolding of these regimes in different places, means there is overlap and different regimes may co-exist in time and space. The concepts employed here, of neostructuralism and retroliberalism especially, are characterisations that we favour in order to document and categorise historical changes in aid. We suggest and discuss four 'aid regimes' in the remainder of this chapter. Firstly, there is the long period of *modernisation* up until about the mid-1980s, though elements of colonial development funding persisted into this period for a number of countries not yet independent and neoliberal influences began to be felt through aid prior to 1990s. *Neoliberalism* then held sway during the late 1980s and 1990s with early structural adjustment programmes and a move away from state-led development. The *neostructural* regime, marked by the launching of the MDGs in 2000 then accounted for the increases in aid in the first decade of the new millennium, whilst the period since the global financial crisis of 2007–08 is what we term a 'retroliberal' phase, still with aid increases but, as we will see, some fundamental shifts in the types and practices of aid. Before we begin this discussion it is necessary to outline the precursors and roots of aid during the colonial period.

Colonialism and 'aid'

The origins of aid are usually seen to lie in the post-1945 environment of decolonisation and the Cold War. We saw in Chapter 1 how President Truman's inaugural address in 1947 laid out the rationale for a new postcolonial development age. However, we can see within the colonial period, and especially in its later years, some important elements that came to characterise the way aid was delivered subsequently and its rationale. We look at these briefly before analysing in more depth the change in subsequent 'aid regimes' that unfolded with decolonisation.

The empires that the major powers established particularly after the mid-nineteenth century redrew the map of the world so that most of the Global South was tied to the North in some form of colonial relationship. Great Britain, France, Germany, Belgium, the Netherlands and Italy[1] had been joined by USA and Japan in claiming territories in Asia, Africa and Oceania. Meanwhile

Austria-Hungary and the former Ottoman Empire maintained a hold on territories in parts of Europe and the Eastern Mediterranean, at least until 1918, and the defeat of the Ottoman-German alliance in the First World War.

Such colonies were claimed and held for a variety of reasons, whether strategic or overtly linked to economic exploitation. 'Development' in terms of improvements in the incomes and welfare of their inhabitants was rarely, if ever, claimed to be a primary justification for the colonies nor a primary responsibility of the colonial powers. To the extent that it was even considered, it was assumed that colonial development, involving the growth of cities, railways and ports and the founding of economic industries (usually in hands of foreigners), would bring peace and, eventually, economic well-being to local people.

That approach changed after 1945. Although independence movements had been present in many colonies prior to the Second World War they were seen as being of relatively minor consequence and it was assumed that colonial rule would persist for many generations into the future. However, the destruction and evident weakness of European powers after the war meant that they could not easily re-establish their control. Local people would not readily accept that colonial rule should return and this was especially the case in Asia, where European control had been temporarily replaced by Japanese occupation. The rise in nationalist movements, pressure from the United States and later the United Nations (which strongly reflected the growing superpower's concerns and agendas) and political realisation at home led many Europeans to accept that the days of the old colonial order were numbered and that preparations for their independence should be made rapidly. Such a change of heart was perhaps best exemplified with regard to Africa by Harold Macmillan's 'winds of change' speech in Cape Town in 1960. India and Pakistan had become independent in 1947, wars of liberation were waged in French Indo-China, Indonesia had dispensed with Dutch rule, and there were major forms of resistance to British colonial rule in Kenya and Malaya. The tensions between the capitalist US-led West and the communist USSR-led East and the ensuing Cold War proved a theatre for the playing out of the postcolonial struggles with both sides supporting different regimes. As a consequence, the resultant 'aid' patterns were strongly ideologically-driven.

Acceptance that colonialism was coming to an end forced colonial powers to consider strategies to prepare the countries for self-rule, for little had been done before 1939 to develop an educated, outward-oriented and experienced bureaucratic or political class that would take on leadership of independent states. Instead, former strategies of divide and rule had often created or deepened schisms and the veneer of nationhood and modernisation was thin indeed. Firstly, there was a need to invest in education to ensure that there were capable people to fill the civil service and an informed public to make good decisions in new democracies. Colonial rulers needed to accelerate infrastructural development so that national integration was more of a priority than just providing quick and efficient routes to get commodities to export ports. Urban areas needed to be better planned to cater for rapid expansion, and social services, such as hospitals, could help convince people of the value of an effective government system.

For Britain, the post-war colonial development and welfare legislation was critical in this regard because it defined a new rationale for the late colonial period (Cowen 1984). There was a large expansion in colonial staff as the functions of the colonial state grew and extended to many more parts of the colonies. This approach, for the British at least, was partly inspired by post-war Fabian socialism which argued for much more of a role for the state in providing for the welfare of its citizens and this reasoning extended to the colonies.

In this environment, the funding of such 'development' came from government coffers, partly the operational budgets of the colonies themselves, but partly also from the grants and loans provided by the colonial powers. These were not labelled as 'aid' or 'ODA', but in many cases they acted as this: transfers of resources from the metropolitan power to territories where they were spent on welfare and development. In terms of the mode of delivery, such 'aid' went directly through government accounts, either as recurrent expenditure (for running schools or hospitals) or as capital expenditure (building roads or irrigation schemes). Thus, we can see that colonial development involved a transfer of funds directly into the budgets of the countries – what we would later call general budget support (GBS). This is an important point to note for it marks the early recognition that important long-term development activities need to be controlled locally by a capable and legitimate government agency.

The colonial development and welfare approach had little time to become embedded and function effectively. Instead, the granting of independence meant that colonial development budgets with their accounting within a colonial framework became the basis for new 'aid' budgets and flows. Consequently, what had been metropoles and colonies now became aid donors and recipients respectively. Furthermore, although the timing of decolonisation was often abrupt, the flow of development funds did not end; with independence they often increased. And this established a pattern of aid flows that has persisted: former colonial powers generally became the largest aid donors in former colonies and the geography of colonialism was reworked and re-spatialised to become the geography of aid.

'Aid regimes': concept and outline

Following independence and the establishment of aid flows we can then observe a succession of broadly unified approaches to the conceptualisation and application of aid. A broad term we use to describe such entities is 'aid regimes'. An aid regime exists where there is a general consistency in the practice of aid relations and delivery, as well as a consensus of some nature (even if it is not explicitly expressed) in the motivations and strategies behind the patterns that we observe. In other words, a regime is similar to the concept of a paradigm where philosophy and practice are relatively consistent.[2] Whilst many different paradigms might exist at any one time, the term 'regime' is used here to refer to that which is dominant at the global scale. In reality, such regimes also vary over time and space. First and foremost, we are referring to aid as practiced largely through the ODA mechanism: the role of the USSR and its satellites in the 1950s–1980s, and the People's Republic of China (PRC) after 2000 could arguably be considered very different sets of 'regimes'. Furthermore, change between one regime and another, just as paradigms in development or science shift, is an uncertain affair. We may be able to observe general conditions that lead to the superseding of one by another and we may identify broad patterns as a result, but in reality, the lines are blurred and regimes can co-exist. Notwithstanding these shortcomings, the use of the regime concept is useful and provides a framework for analysis, helping us understand why the direction, volumes, delivery mechanisms and impacts of aid shifts between periods. Elsewhere we have defined aid regimes as a dominant and widely accepted view which: 'conceptualises and delivers official development assistance and is characterised by a general discourse manifested in a set of guiding principles aimed towards broad goals, combined with regulatory mechanisms which deliver certain objectives' (Overton *et al.* 2019: 29).

In the following section we propose and discuss four broad successive aid regimes:

- Modernisation aid regime (c1945–c1980)
- Neoliberal aid regime (*c*.1980–*c*.2000)
- Neostructural aid regime (*c*.2000–*c*.2010)
- Retroliberal aid regime (*c*.2010–present)

The first three regimes above are relatively widely accepted interpretations (especially the first two); the third draws on a term used often in relation to especially Latin American strategies (Leiva 2008), whilst the fourth is a concept we have evolved ourselves in order to describe and explain recent trends and patterns (see Murray and Overton 2016; Overton and Murray 2018; Mawdsley *et al.* 2018). For the purposes of comparison, we discuss each aid regime under the following headings: background, theoretical underpinnings, key agencies, policies and practices, modalities, and effects.

Modernisation aid regime (c.1945–c.1980)

Background

Early approaches to international development assistance spanned a long period, overlapping with widespread decolonisation and the optimistic early days of newly independent states. It covered a period from roughly 1945 (though for many recipient countries it began rather later than this as they waited for independence) until the early 1980s when neoliberal policies began to take hold worldwide.

Decolonisation was critical in shaping early aid strategies. The partition and independence of India and Pakistan in 1947 (and soon after Burma and Ceylon/Sri Lanka) marked the first steps in what became a rapid and accelerating process of the breakup of old European colonial empires through the 1950s and 1960s. Newly independent states were keen to institute new and progressive policies to undertake rapid economic growth and modernisation. This would demonstrate a break from the oppressive colonial past and illustrate to a recently enfranchised population that their new leaders could deliver the benefits of modernity and prosperity. These new states saw development as one of their main responsibilities and priorities. And from the perspective of donors, particularly former colonial powers, it was important that these new states not fail. They wanted to ensure that independent countries in Asia, Africa and Oceania could adopt economic development paths that matched the models of the West and linked them together in new, postcolonial, ways. Williams (2012) has termed this era the 'sovereign order', a period when national states, both donor and recipient, were central to development.

The Cold War then shaped the way donors approached these countries with some enthusiasm. The Soviet Union and, to a lesser extent, China were keen to gain allies in the world and felt they had socialist models of development that could provide a blueprint for independent countries. Although vastly different in terms of control of the development process, socialist development models still promoted forms of change with industrialisation and modernisation, and aid was a key instrument in putting these new strategies in place. Tanzania and Cuba especially turned to these socialist models and received a sharp response and isolation from Western powers as a result. Western donors, especially USA, did not want communism to spread – it was a threat to American business and trading interests and to their vision of an American-centred new world order (as articulated in the inaugural speech of President Truman – see Chapter 1). Western donors, although talking much about aid for development and progress, were quite explicit about their expectation that aid recipients would align with Western geopolitical priorities.

A final background element of this early phase of aid was a strong belief in the power of science and technology. Technology and economic power had won the world war (jet aircraft, atomic bombs, radar and landing craft); now they could tackle poverty. This amounted to a belief that poverty was fundamentally a matter of a lack of resources and that it was amenable to a technocratic solution. Furthermore, there was a strong emphasis on the building industry and infrastructure to lead overall development. This built from the example of the Marshall Plan for the reconstruction of war-torn Europe funded by USA from 1948. Generous funds were made available and Europe soon rebuilt its key infrastructure and reconstructed and modernised its key industries. This success helped keep Western European countries aligned with USA and linked through growing productivity and trade. The Marshall Plan provided a template for international development aid: substantial aid funds could work well when devoted to establishing (or re-establishing) an industrial base for an economy so that it could grow and trade and become more closely politically aligned. Furthermore, Western science and technology, economic models and systems of government and administration constituted the templates for development. Indeed, these notions formed the basis of the Colombo Plan which also served an explicitly Cold War influenced purpose (see Box 4.1).

Box 4.1 The Colombo Plan 1950–77

In 1950, a number of countries in the British Commonwealth agreed to launch 'The Colombo Plan for Co-operative Economic and Social Development in Asia and the Pacific', what is commonly referred to as the Colombo Plan. Although it was joined later by USA and Japan, the Plan had a strong British core with founding countries being India, Pakistan, Ceylon (Sri Lanka), Australia, New Zealand and the United Kingdom. It also had an implicit goal to promote development as an antidote to the perceived growth of communist movements in Asia. Over time, it was joined by countries that became independent in the 1950s and 1960s: Indonesia, South Korea, Malaysia, Myanmar, Nepal, Philippines, Thailand and Singapore.

Although there was no overall blueprint plan for the operation of the Colombo Plan and each country drew up its own strategic development plan and provide much of the finance. However, in practice, the Plan in its first two decades had a strong technical assistance core. 'Experts' were sent to Asian countries to advise on a range of projects from agricultural research, health facilities, to language training. And many students were sent to universities and other training institutions in Australia, New Zealand, UK and USA.

Between 1950 and 1965, the Colombo Plan funded over 34,000 students to study in Australia, UK, Canada, USA, Japan and New Zealand; over 6,200 experts were sent on assignments from those same countries; and just under £Aust 9.2 million was spent on capital assistance (Oakman 2010:82). USA was by far the largest donor over that period, providing over 90 per cent of the capital assistance and about half of the scholarships and experts.

In effect, the Plan was designed as a sort of Southeast Asian equivalent of the Marshall Plan, without the heavy element of industrial reconstruction, but with a similar political objective to build and economic and political order that was closely aligned with the West (Oakman 2000). It also had a strong modernisation ethos: Western expertise, technology and training was to provide the basis for economic expansion in Asia.

Another major element of the Plan – and perhaps its most important legacy – was the way it aligned the emerging elites of the region with Western values, Western ways of doing things, and Western networks. Friendships and loyalties were established. In subsequent years, many political, bureaucratic and business leaders in Asia would look back, often fondly, to their time in Australia, USA and UK, and seek to maintain and strengthen ties there. It illustrated the value of aid, and specific mechanisms such as scholarships to study abroad, in promoting the long-term interests of donors.

Although the Colombo Plan was designed originally as having a short life-span, and its heyday was in the 1950s and 1960s, it still survives today. In 1977 it adopted a new constitution and a new name, 'The Colombo Plan for Cooperative Economic and Social Development in Asia and the Pacific'. Its membership has expanded to 27 but it retains a strong Asian focus. In an allied initiative, Australia even launched in 2014 its own 'New Colombo Plan' 'to lift knowledge of the Indo-Pacific in Australia by supporting Australian undergraduates to study and undertake internships in the regions' (DFAT n.d.). Interestingly, this reverses the flow of students between Australia and Asia (though many still come to Australia on other schemes), but it maintains the aim of building long-term understandings and relationships and align new generations of leaders with Australia's broader interests in the region.

Modernisation was a 'think-big' and technocratic approach: if sufficient funds were made available on a large-scale and modern ideas applied, then poverty could be eliminated. In the 1960s the Green Revolution seemed to vindicate this technocratic approach. Mounting food shortages and the threat of large-scale starvation were averted by the development of new high yielding varieties of rice, maize and wheat which responded well to the application of new (oil-based) fertilisers and controlled irrigation so that two or more crops a year replaced single crops. Technology, it seemed, could control the vagaries and constraints of nature and bring prosperity and security for all.

Theoretical underpinnings

Decolonisation, the Cold War, and a strong faith that poverty could be 'fixed' by modern technology and large-scale economic investment provided the context for the early evolution of aid programmes but they were given particular shape and direction by certain theories of development at the time. These theories had several elements: neoclassical economics, Keynesian economics, structuralism and the range of modernisation theories.

Neoclassical economic theories were in vogue in the post-war era. They harked back to their classical origins, particularly the free market views of Adam Smith and the free trade ideas of David Ricardo. Yet these were tempered with a recognition that market failures could occur. It was accepted that forms of state regulation and intervention could be necessary, though mostly it was believed that the state should stay away from direct economic production. A new branch of economics – development economics – emerged during the 1950s and focused on how these different approaches, based on fundamental principles of economics, could be applied to the context and problems of the 'Third World' (Hirschman 1958; Todaro and Smith 2015). For example, writers such as W. Arthur Lewis examined the dynamics of growth in economies that had a dual economic structure (a small emergent urban industrial sector and a large and stagnant rural sector) (Lewis 1954).

The experience of the Great Depression had given credence to the theories of John Manyard Keynes. Keynes believed that states could and should intervene in the economy through fiscal policies (raising income and spending in strategic ways) to avoid wild swings in economic fortunes: recession could be avoided by judicious increases in public expenditure when the free market was on a downward spiral. This Keynesian justification for state spending could be extended to suggest that the state could manage the economy, through stimulating economic activity through spending, and ensure full employment and, indeed 'development'. It seemed to be an economic theory that was ready-made for newly independent governments: they could dispense with old *laissez-faire* economics and take an active role in economic development.

Taking the Keynesian state intervention line further were the 'structuralist' ideas emanating particularly from the UN Economic Commission for Latin America (ECLAC) based in Chile and led by the economist Raul Prebisch (1962). Prebisch and others developed an argument that suggested that peripheral, commodity-exporting countries, were particularly vulnerable to the vicissitudes of the global economy and faced a long-term decline in the terms of trade vis-à-vis the manufactured products they were locked into importing. Instead of open free trade, they suggested a model of import substitution industrialisation (ISI) which involved states imposing trade barriers (in the forms of tariffs, quota and currency intervention) so that local 'infant' industries could develop and eventually compete on the world market. This saw self-reliance as a virtue and suggested that development should be oriented more inwardly (*desarollo hacia adentro*) rather than through unregulated linking to the global economy through free trade (*desarollo hacia afuera*) (Kay 2011).

As well as the economists, other social scientists had much advice to offer government development strategies and aid programmes. They looked to the way the West had 'developed' and become 'modern' (even if both terms could be open to question). They suggested that change in the Third World should follow these trajectories and they could do so more rapidly. There were clear economic paths to follow: the widespread development of markets, industrialisation, specialisation, foreign trade and urbanisation. W.W. Rostow went as far as to suggest optimistically that if certain pre-conditions could be met (the development of leading sectors, a rise in the savings ratio etc.), then developing states could achieve an economic 'take off' setting a course to continuing and irreversible economic growth (Rostow 1959). Other writers from sociology – Neil Smelser and Talcott Parsons – examined the social and cultural changes that accompanied this process. There was to be a move to individualism away from communalism, the development of urban societies replacing rural kinship structures, merit-based status by achievement would replace (inherited) 'status by ascription', and attitudes would change so that the accumulation of wealth was pursued and prized. The final element of modernisation theory was that key institution in the process of change was the emergence of the nation-state. The state would embody the will and identity of the population and lead and manage the process of development. In time a modern, prosperous and rational society would replace old communal and traditional ways that were seen to be trapping people in conditions of backwardness and poverty.

In all, these theories provided a model for development, the achievement of which could be supported and accelerated by international aid. For economists, aid could compensate for shortages of domestic savings and investment funds, and government taxes, and thus provide the resources to undertake large-scale and rapid infrastructural development. 'Aid' and 'development' became largely uncontested terms with connotations of 'generalised goodness' and 'progress'. This was shared across the political spectrum and was manifested in grand visions such as the United Nation's declaration of the 1960s as the 'Decade for Development'. Thus, despite the range of theories – and the emergence of some radical critiques in the form of dependency theory – there seemed to be a strong consensus over several decades that development was needed, it could be achieved by using largely technocratic and economic methods, and aid was a critical tool for achieving it.

Key agencies

The evolution of the aid environment in the 1950s and 1960s began to see the emergence of certain institutions that took the lead in negotiating aid relationships, developing policies and dispersing aid funds. On the recipient side, the nation-state was at the centre. Here the old colonial administrations, which had typically been minimal in size and limited in scope, had to be greatly enlarged and broadened. Governments would take the lead in planning and managing development projects. Yet this required many extra resources. More educated and trained staff were needed, not only bureaucrats to manage the projects, but also research scientists to improve agricultural production, engineers to design roads and irrigation schemes, accountants to keep the books, district staff to work with local communities and so on.

On the donor side, particular government agencies took responsibility for new aid programmes. For European donors, this often involved a transfer of some functions and staff from old colonial departments to expanded 'external relations' ministries. Yet there were also instances of separate aid agencies being founded, most notably with President Kennedy's launch of an integrated United States Agency for International Development (USAID) in 1961. Increasingly, albeit gradually, specialist and experienced aid personnel were to be found within donor agencies, replacing temporary assignments of diplomats and general administrators.

Multilateral agencies were also important. The Bretton Woods Institutions (the World Bank and the International Monetary Fund (IMF)) had only been agreed to in 1944 and had much work to do firstly in attending to the recovery of war-torn Europe. The World Bank, in particular, set about providing substantial development loans to recipient governments to undertake large infrastructure projects. The United Nations too – founded in 1945 – soon established a number of sub-agencies. Some of these – the United Nations Development Program (UNDP) and United Nations Educational, Scientific and Cultural Organisation (UNESCO) – had development as a core function.

At this time civil society was seen to be in its infancy and incapable of taking a leading role. Development NGOs may have been emerging in donor countries, raising funds for their own projects overseas and development awareness more generally but within recipient states, they were typically small in number and size.

Overall, then, the early modernisation phase of aid was strongly state-centred on both sides. States took the lead in promoting development and in negotiating and channelling aid into projects and state budgets. Given this state-focused leadership it is not surprising to note that many of the key relationships between donors and recipients built upon former colonial systems and personnel. Old colonial powers did not quickly disappear. Instead they maintained links and used aid as a way of cementing new postcolonial relationships. In this way the geography of aid was drawn anew but based on a colonial template.

Policies and practices

'Aid for development' was the hallmark of the modernisation aid regime. There was strong emphasis on both infrastructure (roads, ports, energy) and welfare (education and health). Infrastructure provided the means for economy to grow and welfare gave people the attitudes and skills to become 'modern'. Aid assistance was crucial because of the scarcity of local capital.

A major strategy of aid was to build economic (and political) integration. Transport infrastructure could link economic sectors nationally, for example by connecting rural food production locales with urban food and beverage processing industries. In terms of aid for industrial development, a number of structuralist-inspired nationalised utilities and transport industries were supported (such as railways, telecommunications, energy etc.) although aid projects were generally limited in terms of direct support for nationalised industries.

Although modernisation theory pointed explicitly to urban industrial growth as the driving force for development, much aid in practice was devoted to the rural sector. This was partly because, until the 1970s (and beyond), most people in developing countries still lived in the countryside and depended on agriculture. Accordingly, a focus on improving rural livelihoods steered many development projects: irrigation schemes, rural electrification, extension services, rural schools and health centres, etc. As well as improving the welfare and incomes of rural people, they were designed to slow the movement of people to cities as rapid and uncontrolled urbanisation posed a problem for new states. Rural development was thus intended to maintain a social and political spatial order at the national scale.

A final feature of aid policies and practices at this time was the appearance of aid 'conditionalities'. Given the Cold War geopolitical context, donors wanted to link aid donations to some sort of political and diplomatic return. Rejecting Soviet overtures, hosting Western military visits or bases, voting alongside Western powers in the UN or offering favourable terms to Western businesses to invest in country were all signs of the way these implicit conditionalities operated. Terese Hayter (1971), in relation to Latin America, coined the term 'aid as imperialism' as a way of illustrating how the hidden conditions of aid led to two-way flows of resources both to and from the recipients.

Modalities

With policies and strategies referred to above in place, aid was delivered in particular ways and modernisation was associated with particular 'aid modalities', which we define here as the channels and processes involved in getting aid resources from donor to end user.

Firstly, we can note that there was hangover from the colonial era for a number of countries following independence. This occurred when donors, as ex-colonial powers, agreed to continue to fund budget deficits directly from their own coffers and into those of the new government. Later this modality would be termed 'general budget support' yet at this time it was merely an expedient move to support new governments who struggled to meet the gap between limited local sources of revenue and a mounting expenditure bill. Gradually though, such general support was phased out as donors were reluctant to pour untagged funds into a budget over which they had no direct control.

Replacing budget support were projects. Projects involved a particular framing of development activities and aid. They involved discrete activities of a fixed duration with pre-defined goals and outputs. Projects were popular with donors. They were straightforward to manage, they could be planned and there was a clear completion date (even if they were not always achieved on time or within budget). Funds could be allocated and results could be seen and measured directly. In time, projects tended to be 'scaled-up'. Larger and larger schemes could bring benefits to a wider area and replication of project templates in different places could have a similar effect. Dams and irrigation schemes became bigger, energy and roading schemes more ambitious and national projects (universities, tertiary hospitals) more prominent. Also, there was a call to integrate disparate projects more effectively. This led to the concept of, for example, integrated rural development schemes, where various projects (water control, electrification, roads, new seeds and mechanisation, rural markets, and training) were planned and constructed together.

Effects

As aid budgets and management systems grew, recipient countries saw many projects spread over the countryside and towns, and despite an end to colonialism, there seemed to be just as many foreigners involved – as donors, experts, engineers, evaluators and project managers. There were many tangible and physical signs of development and there were some significant improvements in numerous countries in terms of services, infrastructure and welfare. But there were also failures and mounting concerns. One of the main issues was increased debt: development loans, albeit on concessionary aid terms, still had to be repaid.

Whilst these problems led many commentators to point the finger of blame at local institutions, officials and politicians, the donor agencies themselves were also culpable. The aim of winning the allegiance of new states as a priority in the Cold War environment meant aid was maintained even when regimes were seen to be undemocratic or corrupt.

Overall, we can conclude that modernisation was associated with the portrayal of development as a shortage of resources and know-how. This had to come from the West and come in the form of aid. Poverty was the public justification for aid but the political and economic interests of donors were always strong, if often in the shadows.

The face of aid was the big projects (dams, roads, electricity, etc.) and aid aimed to build a modern economy and society using the template of the West.

Neoliberal aid regime (*c*.1980–*c*.2000)

Background

Political-economic and ideological shifts in the West in the late 1970s and 1980s were to have a profound effect on the international aid world in the 1990s. Western donors had pursued a modernisation programme through aid and constructed ways of operating through the 1950s to the 1970s. But abruptly, and without consulting their recipient partners, they were to change these fundamentally. The oil crises of the 1970s had deeply affected Western economies. Facing recession, debt levels had risen and economic inefficiencies had become deeply embedded in their economies. Financial institutions also saw a mounting crisis as developing world borrowers struggled to service their loans.

A political revolution, partly in response to these economic trends discussed above, occurred with the almost simultaneous election of right-wing administrations in the UK (Margaret Thatcher in 1979) and USA (Ronald Reagan in 1980). Both instituted harsh economic reform packages, cutting taxes and public expenditure, deregulating the domestic economy and privatising state-owned enterprises. The original experimental site for neoliberalism had been Chile however, where anti-socialist dictatorship was established through a coup in 1973. The changes in economic policy were also often linked to a new moral conservatism that rejected the view that inequality was a responsibility of the state and instead stressed individualism and enterprise as the way out of poverty. Despite severe economic disruption and rising unemployment, the reforms were pushed forward and this neoliberal approach – promoting the role of market forces ahead of a diminished state – became deeply entrenched in Western economies in the 1980s.

During the 1980s, the Cold War continued to rage and indeed intensified with the USA and UK becoming more assertive against a faltering USSR. In this atmosphere of heightened rivalries and geopolitical paranoia, aid continued – and even increased slightly during the 1980s (Figure 4.1). The West was still very keen to maintain and bolster its base of client states in the developing world. However, the Cold War began to dissolve as the Soviet bloc moved to reform and democratise its systems. With the dramatic fall of the Berlin Wall late in 1989, the old enmities fell also and there was much less need to use aid as a diplomatic tool to keep countries on side. In this environment, aid changed. Firstly, aid levels fell markedly in the early 1990s (Figure 4.1). Secondly, rather than maintaining compliant regimes, aid was used increasingly as a means to export neoliberal policies and the deregulation reforms undertaken in the West in the 1980s spread quickly and profoundly to the developing world after 1990.

Theoretical underpinnings

Neoliberalism may have been backed by politicians but it was foremost an economic theory – some would say an ideology (Harvey 2005). It was associated particularly with the 'Chicago School' of economics led by Milton Friedman and their ideas were disseminated widely at a time when old approaches seemed to be failing. Friedman explicitly rejected the Keynesian fiscal 'demand-side' approach to economic management which used the state to stimulate demand in times of recession. Instead they advocated 'monetarism', using market mechanisms associated with the money supply (the role of market interest rates) to ensure that resources were allocated efficiently and that the correct economic signals were made to ensure necessary changes in investment, spending and employment. It amounted to an essential argument to 'roll back the state': diminish the role of the government in economic regulation and leave market forces to ensure that the economy ran smoothly. This was 'neoliberalism' rather than 'neoclassical' economics because it had a much stronger adherence to the basic economic principles of free markets and free trade and a deeper attachment to moral individualism.

Monetarists believed that states made poor economic decisions because they had to satisfy a political constituency, so key economic policies and institutions (such as national reserve banks which controlled money supply) should be freed from political interference. Economic growth would be the engine of human development and real economic growth was best ensured by letting market forces work freely, not using artificial (state) stimulants. With economic growth, there would be employment generation and, as employment rose, so would wages, and the benefits of growth would 'tickle down' to all in the economy. To neoliberal economists, therefore, the market was the best mechanism to alleviate poverty.

Neoliberalism argued that the fundamentals of economic policy were the key: if the market was functioning properly, then there was no need for governments to have development plans or try and engineer social and welfare policy. There was no particular vision, as with modernisation, of a desired future. The market would take care of everything!

Keynesian intervention was rejected and it was believed that reforms should take place rapidly and without delay. It was conceded that there would be short-term dislocation and some hardship but there was faith that people would respond quickly to the new market environment and confidence and growth would soon return.

The proponents of neoliberalism first set their sights on reform in Western economies but soon turned attention to the global economy together with the question of development and aid. Here there was a portrayal of underdevelopment as the fault of the poor and particularly of their governments and corrupt politicians. Political reform was seen as critical for the developing world. In addition, aid was seen as a contributor to underdevelopment rather than development. According to neoliberals, it helped build and support bloated bureaucracies and interventionist and inefficient economic policies. Therefore, aid had to be cut and/or used as a way of forcing change. Aid could be a means to export the neoliberal recipe to the developing world through imposing new conditions for countries that received assistance. There was a view that 'we' (in the West) had reformed our economies, now we will tell you how – and 'encourage' you – to reform yours in the same manner. In practice also the opening of developing economies would bring great benefits for the capitalist West in terms of locations for investment, markets for exports and sources of natural resources.

Key agencies

Just as neoliberalism changed economic policy radically it also led to a re-drawing of the aid landscape in terms of the institutions involved. Modernisation had put the nation-state at the centre; neoliberalism installed the market in that position. But to institute that change, certain agencies became associated with leading the neoliberal reforms.

The key bilateral donors – the USA and UK in particular – were keen to bring about radical reform in aid but they could not do it alone. They found ready allies in the Bretton Woods Institutions – the IMF and the World Bank – whose boards and senior staff were appointed by the main donors. This small group of agencies, centred on the IMF, the World Bank and the US Treasury became the leaders of change and formed what was known as the 'Washington Consensus' – all had their headquarters close together in Washington DC and all agreed on what changes should take place. The IMF and World Bank were able to take on a lead role because many recipient countries were facing severe debt problems in the 1980s – a result of borrowing heavily in the 1970s when easy credit was available and before the oil price hikes and a recessionary debt crisis took hold. Now they were threatened literally with bankruptcy, unable to service or repay their loans due to the dual effects of high interest rates and a global downturn. Both of Latin America's largest economies, Mexico and Brazil, defaulted in the early 1980s and made this fragility starkly clear. The desperation of borrowing countries who had been lent to under low interest conditions was seized upon and creditors agreed help them out but only if they accepted a strict agenda for reform. These were the infamous 'conditionalities' that forced them to introduce Structural Adjustment Programmes (SAPs – see below). The IMF and World Bank (and other regional development banks such as the Asian Development Bank) were crucial here and they were then supported by the key bilateral donors who reinforced the need for change, using their own pressures and aid policies to force reform.

Whilst the Bretton Woods multilateral agencies were prominent, other multilateral agencies became less prominent, in particular the United Nations. Indeed, we could say that the centre of power shifted from New York (the headquarters of the UN) to Washington DC. The UN after all was not governed by a small number of big donors (as was the World Bank and IMF) but by a much broader base of country votes. Furthermore, UN agencies themselves came under attack. The US Reagan administration withdrew funding from UNESCO when it disagreed with its strategies and leadership, for example. Other UN agencies felt the pressure and gradually fell into line with the reform agenda, though the UNDP did continue to highlight the effect of reforms on increasing levels of poverty through the 1990s.

The Washington Consensus therefore opened the space for the private sector to have a much more prominent role in development processes, not directly in terms of undertaking projects or deciding policy or receiving subsidies, but instead simply being given fewer restrictions and more room to operate freely with minimal regulation. Privatisation also created new opportunities for private companies to operate and provide services that were previously the responsibility of government.

The state, particularly recipient states, on the other hand, faced severe restrictions and enforced downsizing. There were reductions in the functions of the state as state-operated services and industries were privatised, there were large cut-backs in the size of government departments as many public servants lost their jobs, and the ability of the state to regulate the economy was severely curtailed.

By contrast, the space for civil society involvement widened considerably during the neoliberal era (Agg 2006). Donors were reluctant to fund state-run welfare services but this vacuum could be filled by NGOs who were seen to be closely connected to local communities and in touch with their needs. NGOs became, in effect, sub-contractors to the donors, running projects for them on the ground. Their concern for community-level welfare also meant that they could help address some of the casualties of neoliberal reform, those who had lost their jobs or access to government services. The number of development NGOs increased markedly worldwide during the 1990s as did the volume of aid money going to the NGO sector.

Policies and practices

The neoliberal aid regime involved using aid as a tool for economic reform. Aid for development was not a key priority as it was assumed that if the economy functioned well, development would take care of itself. In addition, there was to be a 'short sharp shock': measures would be introduced rapidly and simultaneously. The state would be rolled-back, the market 'liberated' and then economic growth would follow, supposedly bringing benefits for all.

The dominant policy prescription was the package of measures wrapped up in Structural Adjustment Policies (SAPs). SAPs ostensibly involved the same range of measures that some donor countries had instituted themselves in the 1980s though in many cases the changes were brought about more rapidly and cut more deeply into the state apparatus. Furthermore, despite the use of conditionalities to force recipient countries to liberalise and demolish their own trade barriers, many aspects of protectionism remained in the West, notoriously in agriculture, which was seen as perhaps the principal sector for opening-up to the world economy in developing countries given the comparative advantage in this sector in the periphery.

Neoliberalism, through SAPs, had several key elements in terms of economic policy:

- *Deregulation*: remove unnecessary controls on the ways markets operated.
- *Balanced budgets*: in order not to distort money markets, states should not borrow and, thus, their expenditure should not exceed the revenue they could raise. However …
- *Reduced expenditure and taxation*: governments should cut the size of their operations in nearly all areas and reduce direct income taxes substantially so that individuals had the incentive to work and save. On the other hand, new revenue might be raised by the imposition of indirect consumption (value added) taxes.
- *Privatisation*: the state should no longer operate as a provider of economic goods and services. It certainly should not run nationalised industries in fields such as manufacturing. Its operations should be sold off (this would help reduce debt) and turned over to the private sector in sectors such as transport, telecommunications, energy supply and even health and education.
- *User-pays*: rather than services being provided free through taxation revenue by the state, such as health and education, consumers of these services and the myriad of government functions, were asked to pay directly the true cost of providing them.
- *Trade and investment liberalisation*: barriers to trade and investment must be lifted. These included removing tariffs, quota, and other restrictions on imports and investment. This would allow domestic industry and agriculture to compete at the global scale and reduce barriers to capital inflows.

- *Currency and money supply deregulation*: governments should not manipulate the value of interest rates nor the national currency. These should be left to find their own value on international markets.

The SAPs were introduced as the immediate response to perceived economic crisis, typically these were instituted through the later 1980s and early 1990s. As a consequence, aid levels were cut (Figure 4.1). Overall, then, neoliberalism posited the view that a lack development was the result of poor governance – too much interference in the economy, corruption and inefficiency etc. – and a private sector that was too shackled by regulation. Recipient states were savaged by SAPs then blamed for their own inadequacies!

Modalities

We have already seen how this new neoliberal aid regime was put in place. Conditionalities were imposed through SAPs to force change on often reluctant and increasingly desperate recipient states. It was in essence a form of coercion. This was very much a top-down or outside-in imposed form of development and aid was a key vessel bringing the neoliberal changes. In referring to two prominent critics of these neoliberal approaches of aid – Joseph Stiglitz and Jeffrey Sachs, both of whom had worked for the World Bank – Gwynne *et al.* (2014: 5) echo how the IMF used a 'cookie cutter' approach to the reforms it insisted upon. Teams of economists often had little experience of the particular countries that they reported on and 'advised'; rather they applied a simple set of 'magic rules' that were applied to a diverse range of recipient countries.

Although direct budget support had diminished during the modernisation era, this fell out of favour even more as a modality. Projects instead were favoured but these also underwent change. Rather than have large projects managed by government agencies and funded by donors, neoliberalism led to a shift whereby projects were more directly managed by donors themselves or their sub-contractors (often NGOs or private companies). This was the case for projects both managed within donor countries and in-country. Also, there was a tendency to downsize the scale of projects: the age of massive rural development schemes funded by the World Bank, for example, seemed to pass to a situation whereby smaller training schemes (to train civil servants in deregulation or to promote small business development, for example) and local projects (water supplies, sanitation, schools, etc.) were more common. Projects therefore tended to be shorter-term, smaller-scale and more localised. Furthermore, the projects came with new strict sets of conditions, in line with the SAPs so that recipients were tightly tied to donor requirements on how to manage economic activities and public services.

Effects

The effects of neoliberal reform were rapid and severe. The austerity packages with cuts in government services had an immediate impact creating rising unemployment and poverty (Figure 4.2). People had less money in their pockets (high inflation eroded the wages of those who still had jobs) but had to pay more for services. Privatised industries frequently shed staff and raised charges, and businesses which had enjoyed a degree of protection behind tariff walls now faced competition from cheap imports – many failed. These were often urban dwellers and urban poverty levels rose appreciably throughout the developing world. This trend was exacerbated by rising inequality in the countryside due to privatisation of

Figure 4.2 ***Structural adjustment***

Source: Cartoon by Polyp reprinted by kind permission.

landholding systems and the associated rise of export-oriented production. This perpetuated already burgeoning rural-to-urban migration. Furthermore, the poor and unemployed now found that their health centres and schools were charging fees or cutting services.

Alongside a rise in poverty and inequality, the attack on the state meant that some governments struggled to survive. Structural adjustment loans did not seem to succeed in promoting growth or even 'widespread policy improvements' (Easterly 2005: 20), but states had much fewer resources at hand. Pared back to the core, functions, such as law and order, also suffered and weakened states began to face internal threats, such as regional secessionist movements or organised crime. Weak states then faced cumulative effects as over-stretched revenue or customs departments were less able to raise the income streams needed by government. These problems began to mount to the point where even the private sector seemed to be concerned for the survival of the state – it was clearly time to re-assess whether neoliberalism had gone too far.

Despite all these problems, it is important to note that neoliberalism did have some winners and positive effects. Those who were able to pick up privatised industries or partner with new international investors (typically members of a small, local entrepreneurial class) found new opportunities for profit. Employers faced an abundance of labour, fewer restrictions and falling real wages. The end to price controls was good for food producers. Rural dwellers who produced small surpluses for sale saw rising food prices: the profitability of growing and selling staple foods such as maize, milk or rice improved, as did domestic production. Small businesses too could do well as local markets began to become established (though others struggled as unemployment led to a contraction of local demand). Export-oriented firms, often foreign owned, benefited in the newly deregulated economic context which led to lower wages and a more easily exploitable environment. The neoliberal revolution had been sudden and destructive and it also created a more vibrant market environment.

Neostructural aid regime (*c*.2000–*c*.2010)

Background

Criticism of neoliberalism and some of its harshest outcomes was strong among the left-wing from its first application. Not only was it damaging to the already marginalised in the societies where it was applied it also created and perpetuated a dangerous reliance on primary product exports, which did little to reverse to historic position of the economies of the Global South. However, when the criticism came from the cathedrals of neoliberalism themselves there began a turn against the Washington Consensus. Authors such as Joseph Stiglitz for example (see Box 4.2), who had himself served in the World Bank, were critical of the one-size-fits-all policies. They were also quick to point out that the hoped-for gains following the austerity periods of early applications, had not in fact arrived.

The rapid application of SAPs across the developing world had succeeded in ushering in an era with much less trade protectionism and much more foreign investment. The need for urgent reform was succeeded by a desire to consolidate the economic changes by instituting a more facilitative environment for the market to continue to grow. By the mid-1990s, there was a move away from blunt structural adjustment to more emphasis on constructing and strengthening institutions which could ensure political stability, protect property rights and widen citizen participation (Craig and Porter 2006). One of the upsides of this was that it removed tacit, and sometimes direct support, for a number of authoritarian governments across the world that had been protectors of the neoliberal models – 'strong' governments had been deemed necessary to introduce and protect the harsh measures required by SAPs. In the case of Latin America, once structural adjustment was superseded, the reasons for supporting such regimes disappeared. Thus, by the early 1990s all of the Latin American dictatorships had fallen. Eventually centre-left governments came to dominate that continent until approximately 2010 in the so-called Pink-tide. In the Global North too, the end of the Cold War ushered in – for a time at least – a shift to more centre-left governments including the administrations of Bill Clinton (USA 1994–2000) and Tony Blair (1997–2005).

The realisation of the failures of the Washington Consensus and the failings of SAPs, together with the geopolitical moment led to a more holistic approach to development in general. The tackling of poverty became central to developmental efforts and, as we discuss later, Poverty Reduction Strategy Papers replaced SAPs (Craig and Porter 2003). Rather than 'rolling back' the state, there was a clear expression of the need to 'roll out' the state, or at least a new version of the state. There was talk of intervention and 're-regulation' that would help developing economies deal with the new challenges and opportunities of globalisation. Foreign investment and property rights (including 'intellectual property' in the form of copyrights over software, films and music) needed to be protected and the smooth operation of commerce needed effective state institutions in areas such as border control, law and order, and (light) regulation of banking and finance. Doing this with decimated government bureaucracies would be impossible and the follies of previous development policies were recognised. This latter point was not just due to a more positive opinion regarding the state vis-à-vis the pure free market, it was also a result of the New York terrorist attacks of 9/11 which led to the war on terror, involving efforts to prevent state collapse for fear that such places would become havens for extremism.

Combined with geopolitical factors was a more positive and informed public view in the West about the perils of underdevelopment: public campaigns such as *Make Poverty History* and *Live 8* (20 years after *Live Aid*) suggested that there was a significant proportion of the electorate that wished to see aid improved (Action Aid 2008). The lack of development was seen as the result of inadequate support for education and health and the continued centrality of the debt burden (which new social movements such as *Jubilee 2000* campaigned against). As a consequence, there was a shift from blaming the poor to blaming the lack of effective aid to support poverty alleviation. This came in part because of the development of the Millennium Development Goals (MDGs), which galvanised and symbolised global development efforts during this period. The DAC began in the late 1990s, and concretised in the mid-2000s, a shift towards measuring aid progress through a broader range of development indicators and more inclusive measures.

Theoretical underpinnings

The neostructural period, as the name suggests, was underpinned by structuralism, as developed earlier by Raul Prebisch and others, but with significant differences (ECLAC 1990, Ocampo 1993). Towards the end of the neoliberal period in Latin America there was talk of 'neoliberalism with a human face' and a shift to market-based policies that took human welfare more seriously. Yet, in reality it could be suggested that the objective of governments was not so much to reduce poverty as an ideological goal in itself, but to win the compliance of the electorate for market reforms and lubricate the workings of the market. This eventually evolved into what has become known as neostructuralism. This approach makes a clear argument that the market alone will not solve economic distribution problems and will not lead to growth. A society that is more equal, better educated and healthier would help facilitate growth. Some referred to such ideas as 'growth with equity', but really it was more like 'equity for growth'.

The ideas of this period were based on structuralist concepts in a number of ways. The original structuralists believed that the historic insertion of the marginalised economies of the world as primary product suppliers to the global economy largely determined their subsequent low dynamism. Neo-structuralists too were keen to increase the value-added component of exports and move away from reliance on primary products. There was a role for the state in subsidising the attainment of skills and innovation in order for this to happen, but never was there a suggestion of a return to ISI with a heavy-handed state regulation and subsidies. The free market was still at the centre of the policy. However, there was a more explicit recognition and analysis of the failures of the market – in terms of stimulating research and development, in terms of correcting damaging inequality and in terms of protecting the environment. Where such failures existed, intervention was considered justified and positive.

Another way of referring to this collection of ideas – or paradigm – is to think of it as Third Way politics, located somewhere in between capitalism and socialism – this is certainly how it was sold by UK Prime Minister Tony Blair and his academic advisors, such as the renowned sociologist Anthony Giddens. In reality it was more closely associated with neoliberalism, with some adjustment at the margins (Leiva 2008, Murray and Overton 2011a, 2011b). Mixed state-market models became in vogue, characterised by public-private partnerships and other centrist policy tools. There was also an explicit turn to the securitisation of aid allocation – the carving out of the state during the 1980s led governments in the West to view aid allocation with scepticism – stronger states would be required to ensure the objectives were met and funds were not siphoned off by corrupt officials and politicians.

Key agencies

If the key shift between modernisation and neoliberalism in terms of aid could be seen as a move away from New York to Washington – the neostructural period saw a return to New York and the United

Nations, and to Paris with the DAC of the OECD. This was embodied in a number of initiatives which sought to codify and apply moves towards aid effectiveness. This signified a major return to internationalism perhaps best exemplified in the MDGs, launched by the UN, which placed poverty reduction, holistic development and the use of aid in order to achieve those things as central objectives.

At a more operational level, the DAC of the OECD took a lead role in the quest for improved aid effectiveness. Starting in Rome in 2003, it oversaw a string of large international 'High Level Forums' on aid effectiveness. These included Paris in 2005, Accra in 2008 and Busan in 2011. But it was the Paris meeting that was the most important (Box 4.2). That meeting, which included both donor and recipient government representatives, agreed to the adoption of five key principles – and associated targets and indicators – that were to guide aid delivery. In retrospect, these principles should be regarded as revolutionary, especially against the backdrop of previous neoliberal approaches. The first principle of the Paris Declaration – ownership – stated that recipient countries should own their own development. In practice this meant that governments should put in place clear strategies, institutions and funding to pursue development and poverty alleviation. The second principle – alignment – then committed donors to fall in behind these recipient strategies and institutions and augment the funding considerably. This aimed to end practices such as tied aid and allow recipients to manage the development process more directly. The other principles – harmonisation, managing for results and mutual accountability – then fleshed out the way aid should be delivered by calling for more efficient practices and transparent reporting. These principles became part of the mantra of development at the time. Aid agencies throughout the world embraced them and recipient agencies used them to develop their own localised versions and protocols for dealing with donors.

In these modes it is recipient states that are prominent – they must exhibit ownership, and the policies of donors should be aligned behind them in this regard. Therefore, recipient states are seen as crucial delivery agents for development policies established through aid relations. Given the emphasis once again accorded to effective aid, donor agencies were transformed and they were allocated larger budgets. The agencies responsible for overseas development and aid were also in many cases (in UK, Australia and New Zealand for

Box 4.2 The Paris Declaration 2005

The Second High Level Forum on Aid Effectiveness, organised by the OECD, was held in Paris in 2005 (OECD 2008). This built on the foundations established at the first forum in Rome two years earlier. The Paris meeting was pivotal because it brought together a large group of donors and recipients and there was broad agreement as to what comprised 'best practice' in aid effectiveness.

The meeting was attended by a large number of countries and eventually some 137 countries endorsed the declaration, along with 29 multilateral and other development agencies (including the World Bank and the IMF). However only 14 civil society organisations were present at the Paris meeting and their voice was not prominent – something they sought to change, with a little success at subsequent forums in Accra and Busan. The Paris meeting, then was fundamentally about government-to-government relationships and practices.

The key result of the meeting was the 'Paris Declaration', a set of 56 'partnership commitments'. The centrepiece – and what became a mantra for a generation of development practitioners – were the five principles (and 12 associated indicators – see Table 4.1).

Table 4.1 ***The Paris Declaration principles 2005***

Target	*Indicator*
1. OWNERSHIP: Developing countries set their own development strategies, improve their institutions and tackle corruption	1. Countries put in place national development strategies with clear strategic priorities
2. ALIGNMENT: Donor countries and organisations bring their support in line with these strategies and use local systems	2. Countries develop reliable national fiduciary systems or reform programmes to achieve them
	3. Donors align their aid with national priorities and provide the information needed for it to be included in national budgets
	4. Co-ordinated programmes aligned with national development strategies provide support for capacity development
	5a. As their first option, donors use fiduciary systems that already exist in recipient countries

Target	*Indicator*
	5b. As their first option, donors use procurement systems that already exist in recipient countries
	6. Country structures are used to implement aid programmes rather than parallel structures created by donors
	7. Aid is released according to agreed schedules
	8. Bilateral aid is not tied to services supplied by the donor
3. HARMONISATION: Donor countries and organisations co-ordinate their actions, simplify procedures and share information to avoid duplication	9. Aid is provided through harmonised programmes co-ordinated among donors
	10a. Donors conduct their field missions together with recipient countries
	10b. Donors conduct their country analytical work together with recipient countries
4. MANAGING FOR RESULTS: Developing countries and donors focus on producing – and measuring – results	11. Countries have transparent, measurable assessment frameworks to measure progress and assess results
5. MUTUAL ACCOUNTABILITY: Donors and developing countries are accountable for development results	12. Regular reviews assess progress in implementing aid commitments

These amounted to a concise statement of how aid should be practiced and the indicators gave a strong lead to what was considered good and bad aid (e.g. with the use of local fiduciary and procurement systems, untied aid and the recognition of need for assessment and mutual responsibility).

The Paris Declaration received criticism. It focused on state-centred relationships and systems and largely excluded civil society on hand and the private sector on the other. There were doubts expressed about the notion of ownership (Buiter 2007) and there were questions regarding power and the legitimacy of some states to 'own' the development and represent the interests of their systems (Hyden 2008). Alignment and harmonisation seemed worthy in principle but some could argue that practices still put donors in the role of driving the processes and imposing difficult conditions on recipient agencies (Eyben 2007; Wrighton 2010).

On reflection, however, the Paris Declaration did represent a remarkable degree of consensus on good aid practice. Despite the pressures it put on recipients, particularly with the need to comply to external standards regarding financial management and the like, the Declaration also gave recipient agencies a clear statement of some quite firm principles that could be used to hold donors to account. The authors of this volume found, for example, that many Pacific Island government agencies and officials became well-versed in the Paris principles and were able to exercise an assertive approach to donors who they knew had agreed to this virtual code of conduct (Overton *et al.* 2012, 2019).

example) removed from broader departments responsible for foreign relations (where many had been located hitherto) and given a clear mandate to focus on poverty alleviation. These shifts also had an impact on multilateral agencies as well as development banks in an expression of alignment with the MDG-inspired approach which saw wide co-operation between the UN and DAC motivated by the Paris Declaration.

During this period there were attempts to increase the size of the DAC donor 'club' and incorporate the voices of civil society and the private sector. The Busan summit of 2011 in particular sought to bring a number of non-traditional donors (especially China) into the aid effectiveness agenda (Glennie 2011) – although this was met with mixed success (laying the seeds for the emerging retroliberal regime at that time).

The shifts outlined above had an interesting and somewhat counterintuitive impact on NGOs. Recall that NGOs had been central in the neoliberal period often undertaking the tasks that were performed by the state previously. During the neostructural phase the kinds of change that NGOs had historically been championing were made more visible – health, education, and advocacy across a wide range of social and cultural issues. In this sense there was a donor NGO-consensus with regard to what was considered important during this period. However, given the increased centrality placed on nation states, the principal axis of interaction was indeed state-to-state. Therefore, despite the fact that NGOs had more funding during this period, their influence proportionally was lessened.

It is important to reiterate that whilst the state was to an extent 'rolled-out' during this period the market remained central. In particular, in the context of international relations, free trade deals, enforced through an increasingly powerful WTO, were considered crucial. There was no overt subsidy for private sector activity but the sector would reap the benefits of a general policy shift towards liberalising trade. It was often stated that development would come through embracing globalisation, not resisting it.

Policies and practices

The main policies and practices of this period have been noted above but it is worth re-capping here. SAPs were replaced by PRSPs – which placed poverty alleviation as the key strategic goal of development policy. National development plans, often as reworked Poverty Reduction Strategy Papers (PRSPs), also became a core part of policy sets. This saw a major investment in many departments across state sectors, which would help reconstruct the state following the neoliberal cuts of the 1990s. Unlike the modernisation period however, where the emphasis was very much on infrastructure, this period saw major investment in areas where societal benefits were considered highest over the long term – education and health. There was also more emphasis on cross-cutting issues such as gender and environment which accorded with MDG principles. In the search for globalisation with a human face, there was an associated emphasis on increased trade liberalisation embodied in discussion concerning the Trans-Pacific Partnership for example.

Despite the shift towards ownership of policies by recipients this period still involved extensive use of templates from donor countries – financial management, audits, state-owned enterprises, reporting/transparency, anti-corruption, and staff training. These amounted to new forms of conditionalities (Gould 2005, Eurodad 2008, Molenaers *et al.* 2015). They focused on political and bureaucratic systems – what we have termed 'process' and 'political' conditionalities (Overton *et al.* 2019) – rather economic reform through earlier SAP conditionalities. In this sense the effectiveness period brought a large compliance cost for recipient countries – which in the case of smaller ones could be overwhelming, leading, despite the rhetoric of 'ownership' and 'alignment', to what we have termed elsewhere an 'inverse sovereignty effect' (Murray and Overton

2011b, Overton et al 2012). Nonetheless, these practices had the objective of building the capacity of recipient institutions to manage development activities. Transparency and efficiency also involved donors who placed greater amounts of trust in recipient institutions and leadership. With trust in place, donors could step back, provide substantial funding, but not become involved in the everyday details of project aid programme management.

Modalities

This period saw a large shift away from projects to programmes, and in general to 'higher-order modalities'. In particular, the use of sector wide approaches (SWAps) and GBS became more important (see Chapter 5). This led to the increased importance of robust and transparent financial management systems that could track and analyse the impacts and risk of the higher-level modalities and ensure that increased aid funding went through smoothly, without leakages or inefficiencies, through government channels to reach desired targets. Overall, this led a move to longer-term funding programmes that were multi-year commitments reflecting the ascendency given to the long-term implications of aid donation and the centrality of consistency and reliability.

The centrality of recipient government agencies in the neostructural approach put the dispersion and management of aid funding firmly within the bureaucratic structures of government departments. Long- and medium-term term strategic planning, in the form of national plans or reworked PRSPs, provided the direction for line ministries to develop firm plans and commitments to deliver key services, particularly in the health and education sectors (Wohlgemuth 2006). SWAps were particularly important for facilitating programmes of multi-year expenditure that could co-ordinate different activities focused on achieving MDG-inspired targets. Sound financial management was particularly important for aid donors, for it could re-assure them that their funds, matched by significant local government budget commitments, would reach where they were needed and make a sustained and significant contribution to meeting poverty alleviation-related goals. The move away from discrete projects, however, meant that the visibility of aid funding was greatly lessened. Instead of having concrete projects with demonstrable physical outputs, aid donors had to rely on aggregate data and often less explicit indicators. The display of donor logos and publicised ceremonies to open a health clinic or water scheme were replaced by government-generated data on enrolment rates of girls in primary education, or the number of visits to a trained midwife, or the proportion of people with access to potable water. On one hand the temporal and spatial scale of the potential impact of donor aid was greatly increased – change could be affected over a wide area over a sustained period of time – and the role of local government agencies was expanded (with hopefully an accompanying increase in their public credibility). Yet, on the other, higher donor expenditure was accompanied by a lower profile for, and a diminished ability to control aid dispersal by, donor agencies.

This shift to higher-order modalities meant that projects became relatively less important as a means of aid management. Only where there was lack of trust between the recipient state and the donor would lower-order project modalities be used. Here, local NGOs continued to play important roles, as they had under the earlier neoliberal approaches. Again, they were seen as contractors for aid-funded service delivery, with accompanying contractual obligations to comply with donor financial management and reporting systems. Unlike the higher-order modalities, in these situations, donors could maintain their close association with change on the ground – public signage could continue to publicise their 'brand' alongside local civil society. Yet the efficiency of this approach, in terms of scale of impact, remained limited, and, in terms of relative overhead costs, remained high.

Effects

In some ways the neostructural regime can be considered the golden period of aid. The focus on poverty reduction and other targets embodied in the MDGs created a galvanised global project. The Paris Declaration of 2005 defined specific targets and pathways with respect to the pursuit of effectiveness. And the scaling-up of aid activity created larger budgets to be dispersed and managed. There seemed to be a consensus at all political levels with respect to the usefulness and potential effectiveness of aid, and the public imagination in this regard was positive. The example of growth in China during this period was also instructive – the state-led capitalism that had dominated there since reform in the 1980s, had clearly led to historic reductions in poverty. Although this had nothing to with ODA, it did illustrate the potential of the role of state management in poverty reduction. This led to a large growth in the levels of aid given in both absolute and relative terms and it also led to the growth in institutions that administered them. In terms of achieving the MDGs, there were some significant achievements, especially with regard to poverty reduction in East Asia (though, again, this had little to do with aid) and in funding for education and health programmes throughout the developing world. How much this has to do with the new neostructural consensus is debatable. What is certain is that aid regained some of the respect had lost in the neoliberal 1980s: it was seen as a legitimate foreign policy tool as well as a moral obligation of donors.

Retroliberal aid regime (*c*.2008–)

Background

The promise of the neostructural period was soon interrupted by events in the global economy and their geopolitical ramifications. The Global Financial Crisis (GFC) of 2007–2008 represented the largest disruption to international capitalism since the Great Depression of the early 1930s (van Apeldoorn *et al.* 2012). At the time it was considered feasible that the entire international economy might collapse and globalisation was under threat (Wade 2010). Thinkers on the left see the roots of this catastrophe as the over-extension of free market principles during the neoliberal period, particularly in the financial sector, which had become so reckless in its lending and obsession with financial instruments that the whole economy had become dislocated from 'production' (Peck *et al.* 2010; Peck 2010). This manifested in terms of property price bubbles across the world. The networked nature of the global economy meant that the negative consequences of this were soon transmitted in a contagion-like effect. Furthermore, the crisis in capitalism forced a rethinking of 'development' (Hart 2010).

The solution to this problem was not, as some may have hoped for, a better regulatory system of the financial sector and a closer alignment with the 'real economy'. Nor did it lead to punishment for those who had created the problem – the financial institutions. Rather, the opposite occurred – large companies were bailed out and considered too large to fail. The impacts of the crisis had a real implication for economic dynamism in the Global North, where recession took hold in the majority of economies – particularly those that were more integrated into the global financial system.

This had a knock-on effect in the Global South as exports to European and North American economies were reduced. Economies which relied on trade with these regions suffered markedly, yet others were relatively unscathed. The demand for natural resources continued unabated in China and economies, such as Chile, that specialised in production of primary products actually saw a rise in the relative price of such products, leading to gains in the terms of trade, export earnings, and ultimately protection from recession.

Recession in the Global North would have been expected to lead to cuts in aid spending as governments sought to reduce expenditure and prioritise their own economies. Yet, paradoxically, this did not happen to a large extent, and some short-term cuts in aid (and major cuts by particularly hard-hit economies, such as Spain) were soon followed by a global increase in aid after 2008 (see Figure 1.1) (Harman and Williams 2014). Adopting Keynesian-style stimulus packages to their own economies, through bail-outs and the like, some donors such as the UK saw opportunities to use their aid budgets not only to help their trade partners but also to support their own companies operating in the Global South. It has been argued that this period in some ways represented a policy of 'exporting stimulus' – or a shift to policies that would yield domestic growth through the mechanism of 'aid' (Mawdsley *et al.* 2018). Infrastructural projects became popular again (Figure 4.3). There was a clear shift to aid strategies that would return greater explicit benefits to donor countries and companies through which they increasingly did their aid business. This accorded with the emerging slogan of 'shared prosperity' by the World Bank and represented a renewed faith in the global economy to drive a path out of recession. As a consequence of the GFC in the Global North there was a shift away from the centre-left, or Third Way consensus that had been influential in the 2000s. Centre-right governments including those led by David Cameron (UK), Angela Merkel (Germany) and Sebastiáan Piñera (Chile) among others took power. As we will see however, the shift to such politics has had significant consequences for the nature and principles at work in the aid world.

Figure 4.3 ***New road project in Kenya***

Photo: John Overton

Theoretical underpinnings

The concept of retroliberalism is devoid of any particular theoretical underpinning, but we have coined this term to represent a new manifestation of economic liberalism that has marked similarities with earlier modernist state-market alignments. It borrows and cobbles together elements from a range of approaches that are essentially populist in their outcome; short-term policies of growth designed to placate the corporate elite and obfuscate understanding in the electorate. It is pro-business and, in some senses, has a neoliberal base in that trade liberalisation is considered important – although later retroliberal forms such as that we are witnessing in the USA at present are in fact more protectionist and nationalistic in their orientation. At the same time as placing the market in an important position, retroliberal ideas also employ some concepts from Keynesianism, structuralism and modernisation – particularly through the stimulus packages released after the financial crisis, and in the emphasis placed on both infrastructural development and increased trade. In this sense we label it retroliberal as it harks back to the modernist period whilst placing capitalist accumulation at the core.

In as far as retroliberalism may have any theoretical core, we might point to two themes that link it to some wider trends in contemporary globalisation:

(1) Financialisation: there has been a growing interest in discussion on social service provision in the possibilities of attracting new forms of capital. To a large extent, this has involved the use of financial instruments to encourage private sector investors to fund forms of social development. Thus, for example in the area of housing for the poor in some Western settings, new types of bonds have attracted new funding streams that shift some forms of risk away from the state (Baker *et al.* 2019). In Chapter 5 we will see how new financial products such as social impacts bonds, are beginning to appear in the development field. This involvement of the private sector and marketised financial products can be seen as central to a retroliberal approach to aid.
(2) Securitisation: since the end of the Cold War there has been an evident increase in the way aid is being influenced by the wider security concerns of Western powers, particularly in regions where they have been involved in military interventions (Brown and Grävingholt 2016). This was seen in the way large amounts of aid has followed the military in places, such as Afghanistan and Iraq. Furthermore, such aid is usually guided not by straightforward development and welfare goals but more by the need to secure local social and political stability, so much aid is directed at (re)constructing institutions of government and order. However, rather than the 'securitisation of development', some have suggested that we are seeing the 'developmentalisation of security' (Pugh *et al.* 2013). Securitisation is not necessarily seen solely in the retroliberal approach – it certainly underpinned the earlier neostructural approach in the wake of the 9/11 attacks (Howell and Lind 2008) – but the linking of aid with military intervention and security concerns (particularly refugee crises) is certainly consistent with approaches to aid in the past decade and the explicit recognition of donor self-interest.

As a consequence of the above there has been a shift in the development rhetoric emanating from the Global North. The poverty alleviation aspirations of the neostructural period have been replaced by the new mantra of 'shared prosperity'. This imagines that aid can and should be used to promote economic growth and that such growth can be promoted in both donor and recipient countries and enterprises: supposedly, everyone gains from the growth of the global economy and aid budgets can be used to stimulate this. This discursive tool of 'shared prosperity' allows governments to continue to convince the voting public of the need for aid programmes whilst at the same time using those programmes to further the interests and profits of donor companies. In this context, more dynamic sectors of 'emerging' economies have been courted and links established. The imagery of aid has changed as well: images of malnourished children or poverty-stricken schools, for example, have been replaced by pictures which depict thriving businesses that are export-oriented. Yet alongside these shifts, there has been continued prominence of humanitarian and emergency relief. The narrative of 'Third World' disasters requiring developed world assistance helps maintain the notion of 'aid' – and aid budgets.

Key agencies

As a consequence of the rise of retroliberalism there has been a retreat away from internationalism. The Busan High Level Forum of 2011 failed to bring China on board with the aid effectiveness agenda and, by then, most donor governments had moved away from the neostructural agenda. Busan was to be the last of such forums, and the mention and use of the concepts from the Paris Declaration that were so influential in the 2000s has faded. Instead, donors have pursued their own strategies, often reverting back to situations where they are competing with one another and duplicating their efforts. As there is no theoretical or even practical core to the collection of policies that now dominates, this fragmentation should come as no surprise. There has also been an apparent retreat from multilateralism and both the IMF and World Bank have struggled to adjust to a new world order in which emerging economies are challenging the hegemony of established Western donors (Vestergaard and Wade 2014).

One exception to the lack of international co-operation occurred in 2015 with a summit in Addis Ababa in Ethiopia on financing for development, but this meeting was illustrative of how retroliberalism was reshaping thinking about aid. The resultant agreement – the Addis Ababa Action Agenda – made some acknowledgement of the need to address poverty alleviation and achieve sustainable development, but its real focus was on how sources of finance for development could be mobilised (see Box 4.3). And here, the role of the private sector was emphasised together with aspects such as international tax co-operation, the spread of capital markets and international financial and trade flows.

Box 4.3 The Addis Ababa Action Agenda 2015

In 2015 a major international conference took place in Addis Ababa, Ethiopia: 'The Addis Ababa Action Agenda of the Third International Conference on Financing for Development 2015'. It was convened by the United Nations and a large number of countries were represented along with several prominent international organisations. This was the third conference of its type focusing on financing for development. Earlier conferences at Monterrey (2002) and Doha (2008) had achieved some degree of agreement, the Monterrey meeting in particular being able to generate commitments from donors to increase funding directed at the MDGs.

The Addis Ababa Action Agenda, the principal outcome of the meeting, represented a shift in global thinking on development, however. Rather than just focus on aid, the Agenda recognised that a more holistic approach was necessary, given that private resource flows far exceeded aid volumes and that domestic funding sources had to be considerably mobilised. It had links to the SDGs, also launched that year, though these remained rather vague. Overall, the meeting recognised the huge scale of the need to increase financing for developing economies, requiring trillions rather than billions of dollars each year to be found if the SDGs were to be achieved. Infrastructure was given particular emphasis.

What was notable with the Addis Ababa meeting that marked a change from the past were two key elements. Firstly, in a departure from previous neoliberal framings of development, the meeting saw that not only were domestic governments critical in development (as had been recognised with neostructuralism) but also they needed to increase their revenue base, in other words, tax more so they could provide key public services, such as infrastructure, law and order and welfare services. Secondly, there was much talk of the role of the private sector. Here, the starting point was the acknowledgement of the importance of private capital in domestic investment, but this shifted to endorse mechanisms such as 'blended finance', public-private partnerships (PPPs) and policies to facilitate and 'unlock' the potential of the private sector. In assessing the challenges ahead, the Agenda stated:

> *Solutions can be found, including through strengthening public policies, regulatory frameworks and finance at all levels, unlocking the transformative potential of people and the private sector, and incentivizing changes in financing as well as consumption and production patterns to support sustainable development.*
>
> (United Nations 2015a: 2)

Aid was also discussed and there was a call for donors to re-commit to achieving the 0.7 per cent target but, on balance, the relatively small scale of ODA in overall global financial flows was put into context.

The Addis Ababa Action Agenda, in itself, did not radically change the way aid was defined or practiced. Yet it did help redraw a new global context for aid in which it was deemed acceptable, even desirable, for aid to be used alongside private capital in order to achieve economic growth. Although there was some recognition of the SDGs, the overall tenor of the Agenda was the way financial resources could be mobilised at a much larger scale to promote economic development. ODA could help in this endeavour, either by indirectly strengthening institutions and public services (its traditional role), or by more actively working with the private sector to address the substantial demands for more infrastructure and business development.

One of the early material impacts of the retroliberal shift was the movement of separate donor government aid agencies back into wider ministries that deal with foreign affairs and trade more generally.[3] The separate aid and poverty alleviation mandates were subsumed again under institutional structures which placed priority on promoting (donor) trade and strategic interests. It is remarkable how within a few short years of the GFC, a number of very similar moves took place across the world, including in New Zealand, the UK, Australia, Netherlands and Canada (for New Zealand, see Banks *et al.* 2012; McGregor *et al.* 2013). Significant restructuring also took place in Japan, though in different ways (Rocha Menocal *et al.* 2011). This has allowed a much closer alignment of geopolitical strategic and trade-based agendas together with aid (Gulrajani 2017). Aid programmes were often rebranded under a national promotion strategy.

A key change in aid delivery then has become increased prominence of the private sector (Tomlinson 2012; DFID 2011b). Furthermore, corporations in donor countries play a central role in the rolling out of development policy through use as sub-contractors and their involvement in development consortia. In these latter arrangements, the private sector may partner with NGOs, and to an extent the state, but in effect the donor state has become the broker for private sector activity in ways that export stimulus to the recipient economy. The impact of retroliberalism on the nature and activities of the NGO sector has been profound. There was an early attack on NGOs associated with poverty reduction policies. In places such as New Zealand there was virtual stripping of NGO capacity, especially for those who depended on funding from the state. They were forced to realign their activities with the private sector and become involved more with projects to promote business development. In some cases, there were new opportunities for NGOs particularly as providers of networks and local knowledge to help business growth, especially when involved in consortia with donor companies.

As well as the inclusion of the private sector and the repositioning of NGOs, the relationships with recipient states have also been realigned in the retroliberal aid regime. Whereas neoliberalism attempted to downsize the state and reduce its role in aid management, and neostructuralism aimed to rebuild and restructure the state to take a leading role in aid-supported development activities, retroliberalism has involved new demands placed on recipient governments. With the aim of increasing foreign investment, donors have been keen to see states improve the regulatory environment to facilitate and protect private sector growth (Hickey 2013). 'Ex-post' conditionalities have continued (Molenaers *et al.* 2015), rewarding recipients who have undertaken democratic and good governance reforms, such as anti-corruption drives. It has also been observed (Jomo and Chowdhury 2019) that donors such as the World Bank, through their Enabling the Business of Agriculture (EBA) project, have promoted corporate agriculture through reforms of land tenure that may well privatise communal land and encourage the conversion of small-scale farms into large-scale agribusiness development.

Policies and practices

One of the core shifts during the retroliberal period has been the discursive shift away from overt poverty reduction to 'sustainable economic growth' and 'shared prosperity'. This has concrete outcomes in terms of the types of policies that eventuate. In rhetorical terms aid programmes are often sold to the public as being in favour of stimulating smaller businesses in recipient countries – using the imagery and rhetoric of microenterprises or microfinance. In reality, though it is often relatively larger donor businesses that have prospered.

One of the more interesting, and indirect outcomes, of the GFC with regard to aid has been the convergence of Chinese and DAC country aid approaches. Programmes from China, to the extent that we have information on them, have been based on concessional loans, in particular for infrastructural development, and associated particularly with the use of Chinese companies in delivering that assistance. During the neostructural period this may well have, and indeed was, frowned down upon as an example of 'tied aid' or 'boomerang aid' (Figure 4.4). Retroliberal policies however have much in common with these approaches. There is a new overt

Figure 4.4 ***Boomerang aid***

In 1983, *New Internationalist* published on 'boomerang aid' and claimed: '"Tying" aid to exports from donor countries means that seven out of every ten aid dollars come straight back to the donors. This diverts aid from its real purpose – the relief of poverty' (Williams 1983). This theme of aid being used to return benefits to the donor economy – neatly summed up in the *New Internationalist* cartoon – has been a longstanding one in the field of aid and has resurfaced strongly in the past decade.

Drawing by Clive Offley. Reprinted by kind permission of *New Internationalist*. Copyright New Internationalist. www.newint.org.

self-interest and programmes benefiting donor economies abound. One such example is the way donors have spent much more on tertiary scholarships: students from developing countries are supported to study in education institutions in donor countries, their fees and living costs being spent there rather than in their home countries. China has also adopted this approach, no doubt also seeing other benefits in the long-term relationships formed with graduates. Interestingly, in the light of the convergence, we are also seeing the emergence of greater co-operation between traditional donors and China in particular. 'Trilateral aid' approaches, linking established and non-traditional donors, and recipient governments, is becoming more common (McEwen and Mawdsley 2012). This convergence between China and the West in ways of operating has occurred not just for economic reasons: China and the DAC countries compete for influence in Africa, Latin America, Asia and the Pacific. In this regard we are seeing a positioning for international geopolitical influence coming through and this revives a very long-standing motivation for aid that was so apparent during the Cold War.

Another feature of aid policies under retroliberalism, accompanying the shift away from poverty alleviation in developing countries as a core focus, has been the inclusion of more activities within donor countries within the framing of aid. Such expenditure within donor countries has long been an element of aid, particularly through mechanisms such as scholarships offered to students in the Global South to study in the North. It is also seen in the way experts from donor countries are contracted to offer technical advice overseas. In these ways, although some benefits flow to developing countries, usually in the form of longer-term capacity building, the immediate benefit of the aid expenditure is to the donor economy where the bulk of the aid receipts are disbursed. However, in the retroliberal period, we have seen even greater use of these approaches and a significant growth in the inclusion of in-donor country costs borne in accommodating refugees in their first year (see Box 2.4). Such expenditure has become significant, with Germany, for example, devoting some 25 per cent of its ODA budget to refugee costs in Germany (Cheney 2017). Such expenditure is all spent within the donor economy with no apparent benefits to the development and welfare of people elsewhere.

Modalities

The move to retroliberal aid relations has seen a shift away from higher-order programmes back to project-based aid. This has been especially associated with infrastructure and energy projects. Although some SWAps and forms of programme aid have continued due to earlier commitments – and there is still significant funding in the education and health sectors, there has been a move away from SWAps and GBS as donors search for shorter term indications of value for money and, it could be suggested, more direct control over their funds (away from the ownership and alignment principles of the Paris Declaration). The time horizons for evidence of aid impacts has been shortened and based more on economic evidence and physical indicators (length of roads constructed, or number of people trained, for example) rather than more longer-term qualitative indicators (attitudinal shifts that lessen gender-based violence or challenge forms of societal discrimination and exploitation). We have also seen a rise in the sectors most obviously associated with short-term gains (construction, trade, financial services) that has resulted in the evolution of modalities including financialisation of development. During this period there has also been a shift to the inclusion in ODA of costs that are incurred within donor countries, for example the costs of hosting refugees, peacekeeping costs, and subsidies for the private sector.

Effects

We are currently in the middle of the retroliberal aid phase, and as such it is very difficult to analyse outcomes and effects. Legacies of past policies can persist, and without the benefit of a longer run view it is difficult to predict. However, there are some trends that are worth considering.

One of the most surprising outcomes of the post-GFC aid world has been, until recently at least, the maintenance of the size of aid budgets. At first this appeared very difficult to explain. However, as noted above there has been a shift to policies of self-interest beneath a discourse of 'shared prosperity'. This system of exporting stimulus is one of appeasing and redressing the impacts of the GFC in donor country economies. Some of these shifts, it could be argued, are not aid at all! Indeed, the whole concept of ODA has shifted so much, it is sometimes difficult to recognise it.

There has been a clear shift from international co-operation to competition between nations which has seen a convergence of aid policy towards the Chinese model of concessional loans for infrastructure. And, as a consequence of this, we have seen a move away from the aid effectiveness agenda to a more self-interested financialised and private sector-led regime. The impacts of this on aid recipient countries and those for whom aid is intended can only be hypothesised.

Conclusion

The purpose of this chapter was to identify collections of ideas concerning the motivations for and practices of aid policies. We argued that it is possible to identify 'paradigms' in aid philosophy and practice which we termed aid regimes. We noted that such characterisation is imperfect, as aid regimes overlap both across time and place. However, we identified four regimes (modernisation, neoliberalism, neostructuralism and retroliberalism) preceded by a period of colonialism and immediate postcolonialism when the geopolitical antecedents and patterns for contemporary aid patterns were forged.

In Table 4.2 we summarise the aid regimes we have proposed in this chapter. In each we outline the broad dates of the dominance of that regime, the underlying worldview attached to it, some of its major policy outcomes and impacts.

These regimes point to significant consensus, albeit not always explicitly expressed, with regard to the philosophy and resultant policies of aid at various points in time. In some ways the current regime is an anti-regime in that it does not involve an explicit international accord on what aid is or should be and does not proscribe a set of policies to achieve that. Notwithstanding, it is possible to see the emergence of what we might better term a decentred regime. What is of most concern is whether that will lead to a complete dislocation of the original motivations for aid from international practice and the death of aid. Or, in contrast, will it lead to a galvanising among those that would like to see at least a partial return to the more progressive objectives of the past – such as poverty reduction and ecological sustainability? This will come as a result of political choices in the Global North, no doubt as is always the case, punctuated by unpredictable critical global geopolitical junctures and crises.

Table 4.2 *Aid regimes: a summary*

	Modernisation	*Neoliberalism*	*Neostructuralism*	*Retroliberalism*
	1945–1980	*1980–2000*	*2000–2010*	*c.2010 to present*
Global Events	- Allied War victory - Truman's four-point programme - Cold War rivalries - Decolonisation - Transformations in China	- Economic crises (oil shocks and debt crises 1970s/80s- share market falls 1987 & 1997)- fall of the USSR	- 9/11 and invasions of Iraq and Afghanistan- period of economic growth	- global financial crisis- the rise of China and other 'emerging powers'
Domestic Political Context in the West	- Cold War politics- Kennedy's Alliance for Progress	- Thatcherism - Reaganomics	- Rise of Tony Blair's New Labour (UK) Clinton's democrats	- Swing back to the right: Cameron, Abbot, Key, Republican control of Senate in US
Principles	- Modernist and traditional structuralist ideas concerning role of industrialisation and backwardness of rural development- Keynesian economics- Geopolitical imperative of preventing domino effect across the Third World- socialist modernities- dependency theorists	- Neoliberal theories and monetarist economics- The state crowds out the private sector and leads to inefficiency and corruption- The market will arrive at Pareto Optimum- Benefits of export growth will trickle down to poor through employment- Comparative advantage and trade liberalisation	- The state tackles social justice but in the context of a globalised economy that remains open- Poverty and inequality are seen as consequences of the market but are responsibilities of the state- Deliver the benefits of globalisation and ensuring its trickle down	- The state exists to facilitate economic growth- The private sector should not be crowded out by the state- The state sponsors and facilitates the private sector- Ricardian comparative advantage coupled with neo-Keynesian economics to stimulate the private sector during recession

Continued

	Modernisation	*Neoliberalism*	*Neostructuralism*	*Retroliberalism*
	1945–1980	*1980–2000*	*2000–2010*	*c.2010 to present*
Development goals	- Grow industrial sector- Promote regional alliances- Promote urbanisation and reduce rural inefficiencies	- Reduce government size- Raise productivity- Stimulate exports- Develop the private sector	- Poverty alleviation- Equality promotion- Aid effectiveness through market mechanisms	- Economic growth- Infrastructural development- Stimulate trade and investment through financing
Aid policies and modalities	- Import substitution Industrialisation- Land reform- Support for state budgets and building state capacity- Colombo Plan	- SAPs: privatisation, hollowing out of the state, reduction in social expenditure- Export-orientation- 'Good governance'- Market-based projects- Use of civil society	- MDGs- National interest and development agenda (formally) separate- Poverty Reduction Strategy Papers and poverty reduction-based projects Sector Wide Approaches (SWAPs) and GBS- Reconstruction of the state for security	- Infrastructure development- Semi-tied aid projects- New (returnable) forms of development financing- Development for diplomacy and the rolling together of national interest and developmentalism- Partial return to project modalities

Source: Adapted from Overton *et al.* (2019)

Summary

- The historical antecedents of today's global aid sector lie in the colonial period and the direct funding of budgets by the colonial powers.
- We can identify paradigms in aid where there is a broad level of consensus regarding the rationale, purpose, policies associated with, and the expected outcomes.
- These broad regimes shift across time and space but, generally speaking, feature commonalities in terms of 'modalities' or mechanisms for delivery and application of aid policies.
- Following the Second World War and the coining of the concept of development there have been four main aid regimes that roughly have replaced each other over time: modernisation; neoliberalism; neostructuralism; and, most recently, retroliberalism.
- Notwithstanding the chronological argument above there is also significant overlap within and between aid regimes. Different aspects can be seen to dominate in different places.
- There is at present a significant consensus with regard to the nature and purpose of aid, which evolved approximately following the global financial crisis.

Discussion questions

- What is an aid regime? How could the concepts of aid regimes be criticised?
- Account the broad shifts between aid regimes that we have witnessed since the Second World War.
- What are modalities and how do they relate to aid regimes?
- What are the similarities between the modernisation and the neoliberal aid regimes?
- To what extent does retroliberalism represent a return to the colonial period preceding the rise of aid? Also discuss how it does not represent such a regression.
- Discuss the retroliberal aid regime, and in the light of the recent rise in nationalist populism, predict what will happen to this way of organising the sector.

Websites

- Devex 'the media platform for the global development community': https://www.devex.com/news
- ODI – the Overseas Development Institute (UK): https://www.odi.org/
- The *New Internationalist* magazine: https://newint.org/

Notes

1 The Spanish and Portuguese empires were of an earlier age founded beginning in the late 1400s and many of their former colonies in Latin America had achieved independence by the early 1800s. However, remnants remained in the early twentieth century, particularly the Portuguese territories of Angola, Mozambique and East Timor. Although power from the 'motherlands' (Spain and Portugal) was no longer existent, the forms of domination and control that those powers had brought to their territories persisted.

2 Bellù (2011) has outlined a framework of aid paradigms, though the 15 he describes are rather more specific policy-based approaches to aid than the broad eras we employ.

3 One exception to the trend to incorporate separate aid agencies into wider foreign affairs and trade ministries was the UK which maintained DFID as an independent government aid agency. However, as this book was going to print in June 2020, the UK government announced that DFID would cease to exist and be folded in to an enlarged Foreign, Commonwealth and Development Office.

Further reading

Booth, D. (2011) 'Aid, institutions and governance: what have we learned?', *Development Policy Review* 29(1), 5–26.

Harman, S. and Williams, D. (2014) 'International development in transition', *International Affairs* 90(4), 925–941.

Mawdsley, E., Murray, W.E., Overton, J., Scheyvens, R. and Banks, G.A. (2018) 'Exporting stimulus and "shared prosperity": Re-inventing aid for a retroliberal era', *Development Policy Review* 36, O25–O43.

Overton, J., Murray, W.E., Prinsen, G., Ulu, A. and Wrighton, N. (2019) *Aid, Ownership and Development: The Inverse Sovereignty Effect in the Pacific Islands.* Routledge, London (Chapter 2).

Williams, D. (2012) *International Development and Global Politics: History, Theory and Practice.* Routledge, London.

5 How is aid delivered?

Learning objectives

This chapter will help readers to:

- Understand the concept of aid modalities or modes of aid delivery
- Appreciate the difference between lower and higher-order aid modalities
- Assess different examples of aid modalities with particular reference to programmes and projects
- Analyse how the application of different aid modalities has shifted across time and space and continues to evolve
- Understand the nature and rationale for humanitarian relief and some of the longer-term challenges associated with it
- Distinguish between various types of higher-order modalities including Sector-Wide Approaches (SWAps), general budget support GBS, debt relief and import support
- Understand the rise of private-sector aid initiatives and associated new modalities, the rationale for these and some of the critiques that have been made

Introduction

The issue of how aid is delivered is a crucial one in terms of addressing key questions such as 'does aid reach those for whom it is intended?'; 'does it pursue and obtain its main objectives?'; and, 'is it used effectively?'. A variety of 'modalities' – modes of delivery – are used in the aid world and these can vary greatly in terms of scale and time frames as well as parties and institutions involved. In this chapter we focus firstly on various aspects of modalities at various levels, placing most emphasis on projects and programmes. Firstly, we consider those that have characterised the neostructural aid regime. Then more recent approaches to delivering aid utilising private sector involvement associated with the retroliberal aid regime are analysed. As we will see these are novel and, as yet, largely unproven.

As a framework for analysing projects and programmes, we adopt a model developed for use by the World Bank (Koeberle *et al.* 2006). This envisages a progression in modalities from projects, through 'pooled projects' and 'basket funds', different types of SWAps to GBS. Movement from lower to higher-order modalities increases impact of aspects such as volume, operational scale and policy – and, ostensibly, development benefit. In what follows, we pay particular attention to 'lower order modalities' (projects) and 'higher-order modalities' (SWAps and GBS) and some of the forms in between.

Lower order modalities: projects

Projects have been the dominant mode in aid practice over the past 50 years or so. Projects have a number of characteristics and advantages. They involve defined objectives, means, outputs and time frames – all of which are articulated at the start of the project cycle in a planning phase. This rests on a key assumption: that development interventions involve processes which are observable, measurable, predictable and controllable through rational management systems. Projects also tend to involve development, which is about 'providing things' (whether physical or technical) that will have clear development benefits. Thus, if there is a problem of lack of adequate water supplies in a rural area, a project would involve, say, the provision of water storage facilities (e.g. a reservoir or tanks) and a reticulation system (e.g. pipes and taps to households). There is a clear problem, a planned intervention, a fixed agenda and a finishing point. Once the storage and reticulation systems have been constructed and are operating, the project is executed and the aid input complete. Donors and managers can then celebrate the opening of the facilities, complete an evaluation report and move on to the next project. It is, in theory, a tidy approach where risks are hopefully identified early on and mitigated accordingly.

The nature and scale of projects

Projects usually occur outside direct recipient government support. Government agencies may certainly be involved in the projects, giving permission, helping to identify the priorities, sometimes managing relations with local institutions, and often having officials involved in follow-up or complementary activities. However, typically projects do not involve the use of substantial recipient government resources. Aid donors provide the majority of financial and other resources. In some cases, for example when recipient states are weak and have little presence in remote areas, aid projects may be implemented almost without government involvement at all. In these circumstances, aid donors – whether government aid agencies or donor non-government organisation (NGOs) – will use NGOs or consultants to manage the projects and liaise with local communities. Donor agencies will keep an eye on projects usually through requiring regular reports from the project managers on the ground and it is not uncommon for aid officials to have to manage a large portfolio of projects from afar, dealing with funding, compliance and progress reports. Thus, although project management can be contracted out, there are still high overhead costs for donors.

Projects can vary greatly in size and scale. Many aid projects involve single activities. These are discrete projects with defined scope and time frame. They may involve the building of a water supply system, a school or a wharf. Such projects may have a major impact at the local scale – providing better sanitation systems or a health centre or access to primary schooling – but they are unlikely to have an impact on a whole country. Others may see a project template replicated across several projects and locations: a donor may develop particular experience and technical expertise in a certain area and be keen for this to be repeated, improving the efficiency and visibility of their aid spending. Thus, we have seen Japanese projects for irrigation and agricultural machinery, or Swedish projects on gender equity awareness. In this way, small projects are 'scaled up' by repeating a proven project approach in different places: what is sometimes called a 'cookie cutter' approach (meaning replicated in a standard way). Yet such projects are still rather limited in their ability to affect a wide area and deeply address issues of long-term policy and institutional change.

Small-scale projects have several advantages, largely because they are often linked to particular local situations and, hopefully, involve local actors and a degree of local ownership. On one hand, smaller projects may be regarded as inefficient because fixed and overhead costs are relatively high and economies of scale are not easily achieved. Yet, following Schumacher (1973), high relative costs may outweigh the negative sides of the high absolute costs that big projects incur with high debt burdens, a loss of community-level control and ownership and the potential for large-scale negative impacts should things go wrong.

Other projects can be large-scale, thereby increasing impact at higher levels. A hydro-electric dam, with associated electricity generation and distribution systems, water distribution and irrigation, and resettlement of affected communities can cost many millions of dollars and span a decade or more. They can achieve economies of scale and cost efficiencies, they can tackle development issues over wider areas and they can link different forms of activities (such as infrastructure). These large-scale projects again have a substantial aid donor component, often in the form of concessional loans, though we may see more of a role for recipient government agencies in managing the new facilities and being responsible for loan servicing and repayment.

Sometimes associated with these larger-scale efforts has been a concern to integrate different projects that complement one another. Lessons learned in the 1960s and 1970s pointed to the fact that a single large project could be limited in its ability to bring about development if it did not recognise the need for changes to occur in related activities. Thus, for example, the introduction of new high-yielding varieties of grain at the heart of the Green Revolution in the 1960s were of little use unless there were improvements in water management (requiring irrigation projects), or better marketing (requiring projects to build new roads, storage depots and processing facilities), or technical advice (needed extension services and farmer education), or equipment (leading to projects to provide agricultural machinery), and so on. These gave rise to the approach known as Integrated Rural Development. They were large and expensive operations that could cover a wide area and involve many different facets of the rural economy. Yet, as a modality for aid, they were still based on a combination of projects, i.e. activities that were pre-defined, closely managed and with a focus on particular outputs.

The project cycle and management

All projects, large and small, share a common feature in terms of design and operation, namely what is termed the 'project cycle'. This

Figure 5.1 ***The project cycle***

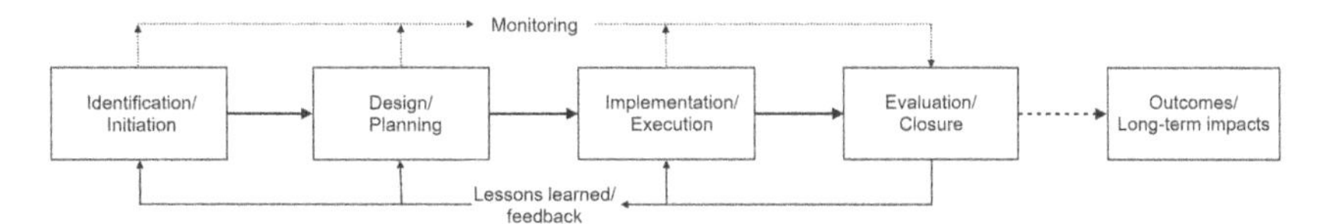

sees a logical progression through various stages of a project: from project identification (where the 'problem' is seen and agreed upon); project design (involving, in turn, the identification of solutions, the planning of interventions and management systems, and, often, aspects such as feasibility or impact studies to gauge the likely success and risks); implementation (the actual expenditure of resources to address the problem and its solution); to project completion, the generation of outputs and the process of evaluation (when the project is deemed to have finished and lessons are drawn from it).[1] The notion of a project cycle then implies that the lessons learned are fed back into the planning and design of similar and future projects. And so the process goes on, presumably with practices improving all the time and aid being used more effectively each time to enhance the development outcomes.

Alongside this notion of a project cycle has developed a particular approach to 'project management'. This is a set of practices and tools that have evolved over time. Project management has become a specialised skill and many practitioners have gained training in this area and been employed widely in the aid world as independent consultants or specialists within aid agencies. Project management, in its increasing sophistication, attempts to ensure that projects are more and more tightly subject to rigorous techniques of prediction, planning, operational control and assessment. Monitoring and reporting are vital components as are various techniques to measure social and environmental impacts. Throughout, projects are seen to be rational and ordered attempts to bring about development, that try to predict impacts, mitigate harmful effects and ensure that aid funds are wisely and efficiently spent. Project success is determined by two main criteria: did the project come within or close to budget? Were the planned outputs constructed or provided as designed? The question of unforeseen impacts or long-terms outcomes of projects are largely left outside of the scope of project cycle management.

The benefits and costs of project modalities

Because they are ordered and in theory predictable, projects have proved to be a very popular modality for aid agencies. Aid donors can keep a close eye on how their money is being spent and there is an end in sight and, as mentioned previously, if things are successfully achieved the positive publicity reflects well on a beneficent donor.

However, projects have fundamental limitations as a modality in terms of aid effectiveness. Their impacts can only be limited to the discrete activities they involve and these are proscribed spatially and temporally. They are unlikely to make profound changes to a nation's well-being, for example, unless they involve literally hundreds and hundreds of separate, but similar, projects in all parts of a country. Furthermore, project management may be effective as an approach to managing a single project but, in sum, is a rather inefficient way of managing change on a large scale. Overheads are replicated and multiplied, economies of scale are not achieved, and it can be difficult to communicate lessons from one project to another. Projects, as we have seen, focus on outputs rather than longer-term outcomes and thus they can miss the need to engage in activities which address the deep-seated attitudinal and structural causes of poverty. They also tend to work on the basis of the certain, the quantifiable, the observable and the manageable. In practice, development processes are highly complex, uncertain and sometimes chaotic and even violent. Projects attempt to contain and manage desired forms of change, whereas local people experience lives which are not so ordered or – by external eyes – rational and predictable. Projects fail when the unknown, the unplanned and the uncomfortable overwhelm the ordered certainty of a project planning document. And finally, projects tend to exclude the local national government. Not only is the opportunity not taken to draw on important government resources and support and develop their capacity, but also projects may act to undermine the perception of the effectiveness and legitimacy of the state by its citizens. If it is aid donors and external agencies – or local NGOs – that build schools or improve roads and water supplies, the state is seen to be marginal and ineffective in development.

Humanitarian relief

Although aid for development is usually channelled through projects or programmes, we should also note that significant amounts of aid are delivered not for explicit development activities but to address pressing humanitarian needs, such as disaster relief or refugee crises. Unlike projects, these forms of assistance are not pre-planned, but arise in response to sudden change: an earthquake, cyclone or tsunami; or a rapid change in refugee movements. Yet, such aid still has much in common with project modalities. Most major donors have developed over time sets of practices, and contingency funds, so that they can respond quickly and effectively to such emergency requests. Thus, once a disaster occurs, donors and many recipient agencies can swing into action, using project-like management techniques. Well-tested logistical networks can be employed, reporting and monitoring templates are used and, hopefully, a fixed time frame is achieved so that the assistance can be seen to come to an end and the crisis responded to.

Box 5.1 Humanitarian relief: the role of NGOs and agencies

The *New Internationalist* magazine in 2018 reported that 136 million people worldwide needed some form of humanitarian aid and that in 2017 $28 billion was spent in the previous year responding to disasters: what amounted to 13 per cent of ODA. ‘Humanitarian aid in this case covers everything from food, cash and the construction of roads or shelter to provision of medicines, vaccines and education for displaced children’ (Perrone and Healy 2018: 16).

The United Nations is prominent in humanitarian operations worldwide. Several of its agencies are engaged in relief work, such as United Nations High Commission for Refugees (UNHCR), United Nations Children’s Fund (UNICEF), United Nations Development Program (UNDP) and World Food Program (WFP), through its Office for the Coordination of Humanitarian Affairs (OCHA) that is responsible for co-ordinating the overall UN response to emergencies. The OCHA manages the UN Central Emergency Response Fund which maintain a large fund available (over $US 400 million in 2019) to be quickly disbursed in response to disasters and humanitarian crises. There is also a very large number of NGOs – perhaps some 2,500 (Perrone and Healy 2018) – engaged in such work. However, NGO work is dominated by several large agencies: Red Cross/Red Crescent, Save the Children, Oxfam, MSF (Médecins sans Frontières) and Cooperative for Assistance and Relief Everywhere (CARE). These organisations are engaged in a range of work, and disaster relief is not necessarily their core operation. However, with their global reach and substantial revenues, they are able to react quickly to disasters and have well-developed systems and experienced staff.

Alongside NGOs and multilateral agencies, most of the larger donor aid agencies are engaged in humanitarian relief. Globally in 2017, the largest such contributor to humanitarian aid was USA with over $6 billion spent. Interestingly, the next largest government donor was Turkey with about $6 billion, but nearly all of this was related to the Syrian conflict and was spent on refugees inside Turkey itself. Much of this cost was donated to Turkey by the EU. Other prominent government contributors were UK, Germany, UAE, Canada, Saudi Arabia and Australia.

Humanitarian relief is needed when disasters strike. Earthquakes, floods, droughts and tsunamis can strike and cause great disruption to people's lives, homes and livelihoods. In these cases, aid can meet immediate needs for food, water, medical supplies and shelter. These require approaches that are not quite classic project modalities, but they do need a quick response to an immediate and clear material set of needs and logistical and other support to help people recover and return to their normal lives.

Unfortunately though, many disasters have human rather than natural causes. War and conflict lead to displacement of people, the loss of homes and often much injury and suffering. In the short term, relief efforts require a similar approach to that for natural disasters and a project-type approach is useful. Materials and supplies are sent to places and communities of need and immediate human needs for survival can be met. However, when the underlying causes remain – continuing conflict and violence – people cannot 'return to normal'. In these circumstances, sadly more common in regions with deep-seated violence and entrenched conflict, aid agencies focused simply on relief and a project approach cannot meet the longer-term needs of affected people. What may be needed is a transition to more development-focused work, helping, for example, to train people in new skills, help them into new livelihood activities, build new institutions or ensure children get education. This, in turn, is likely to call for longer times frames and a shift from a long series of discrete projects to more integrated programmes.

Humanitarian assistance, as with many project approaches, concentrates on delivering resources to places and people in need. It is primarily about the alleviation of immediate suffering, ensuring the recipients have sufficient shelter, food, water and security. It can be effective or not at meeting these pressing needs – and in managing the process of delivery – but rarely does 'relief' turn to addressing what might be the deeper causes of human suffering (be they climate change, political oppression, or human-induced food shortages) or the systems of resilience which allow people to respond to disaster themselves (networks of support, livelihood alternatives, back-up resources). For these important issues, rather different approaches are needed: those which pay attention to long-term and often qualitative changes, and to building up assets and capabilities, rather than meeting deficits. These require an approach which promotes

Figure 5.2 *Distribution of relief supplies in Samoa following the tsunami in 2009*

Photo: John Overton

'development' in different guises, but also a shift in the way is delivered – towards longer-term and broad-based modalities.

Higher-order modalities: programmes

The realisation that projects have limitations in terms of efficiency and effectiveness with regard to large-scale development efforts led to a turn to higher-order modalities, especially since the launch of the Millennium Development Goals (MDGs) in 2000. Even prior to the MDGs, donor agencies, such as the Development Assistance Committee (DAC), had been considering ways to improve the effectiveness of aid and provide evidence for aid producing meaningful results. It required a move from evaluating discrete projects under narrow output-based criteria, to suggesting and assessing ways in which sustained and long-term programmes across many sectors could achieve good development outcomes over perhaps more than one generation. To do this, a crucial change was the need to involve recipient states centrally in the aid efforts not just as passive observers, but as leaders and managers. Such modalities, then, are much more state-managed and involve a direct partnership between donors and recipient government agencies (Reinikka 2008).

The nature and scale of programmes

Higher-order modalities involve a progression over various forms, each of which involves a higher degree of hand-over of control to recipient agencies together with an increase in scale of the activity (Koeberle and Stavreski 2006). Not only does the volume of aid increase, but also the degree of involvement of local institutions increases, yet on the other hand, the degree of overt donor control seems to decline (even though they will have systems in place to ensure they keep watch on disbursement of aid). In short, these progressively higher-order modalities are SWAp and GBS:

(1) *SWAp*: A Sector-Wide Approach (SWAp), as the name suggests, is an approach that involves the whole of a sector (health, education, telecommunications, agriculture etc.) and funding for this. It involves aid donor funds going to the respective government department to be spent on that sector. It requires that the government have a detailed strategic plan for the sector, outlining objectives, targets, strategies and systems that donors agree to, and that they have good mechanisms in place to implement and monitor the activities involved. With a SWAp, donors move to recognise the ability of a given government sector to manage itself. Agreement over the strategic plans and internal management systems are critical: they constitute the basis for a contract between donors and recipients. In many cases SWAps will involve one donor taking a lead, providing funds and some of the technical expertise. Other donors (both bilateral and multilateral agencies though rarely NGOs) will also contribute, whether by adding to a central pool of funding for the department, or quite commonly by nominating a particular part of it to support. SWAps are thus effectively a consortia of donors bringing together several strands of funding to secure a broad-based and continuing source of funding. Under a SWAp, donors provide funds, but substantial funding also comes from the recipient government itself – and again these amounts and how they are shared are specified before the SWAp is signed. Recipient government funding is essential as a sign of their commitment to the strategic plan and also as a basis for future complete self-funding. SWAps mean that the government department

manages all the funds and ensures that the development strategy is pursued. It is a demonstration that they are in control of the development of the sector.

(2) *General Budget Support* (GBS): This is the highest-level modality of aid and one that has sometimes been regarded as the 'Holy Grail' of the aid world. Within GBS, donors have very high levels of trust in the recipient government to 'own' and manage the country's development and are happy to contribute, usually as a coherent group of donors, to the total government budget of a country. GBS involves the highest levels of trust between the government parties concerned. The volumes of aid may be very large and the commitments reliable and multi-year. Aid funds are given to the recipient government and they are fed into and controlled through and by the government's own management systems. However, behind the apparent simple handing over of funds, there is a complex web of agreements and understandings regarding financial management systems, auditing, reporting and consultation. GBS is – or was – regarded by many donors so highly because of its potential for efficiency. Donors do not have to bear the cost of heavy overhead costs monitoring and managing individual projects, activities and flows of funds. There are considerable economies of scale to be had by scaling-up development activities at a national scale. Furthermore, duplications and competition amongst donors should be eliminated – bringing benefits for donors and recipients alike – and local systems are respected and supported (rather than facing parallel and competing donor systems), thus further building their capacity and capability. In these ideal terms, GBS has been regarded by donors such as the World Bank as the goal to which all should aspire for aid delivery.

Although we will see below how new aid modalities have emerged in the past decade, higher-order modalities remain very important. A review by the European Union in 2018 noted that the EU was the world's largest provider of budget support, providing 70 per cent (€1.8 billion in 2017) of the global total (European Commission 2018: 7). The document defined budget support as follows:

> *EU budget support is a means of delivering effective aid and durable results in support of EU partners' reform efforts and the sustainable development goals. It involves (i) dialogue with a partner country to agree on the reforms or development results which budget*

> *support can contribute to; (ii) an assessment of progress achieved; (iii) financial transfers to the treasury account of the partner country once those results have been achieved; and (iv) capacity development support. It is a contract based on a partnership with mutual accountability. (European Commission 2018: i)*

Interestingly, this re-statement of budget support policy put stress on the achievment of the Sustainable Development Goals (SDGs), a clear recognition that long-term and large-scale progress towards meeting the goals will require the active role of states and the use of reliable high-order modalities. The EU sees benefits in this approach in strengthening macroeconomic stability and good governance so that the private sector can thrive. Funding by the EU for budget support has been increasing since 2015, and it amounts to 40 per cent of the EU's assistance to recipient countries. However, despite this commitment in Europe, it appears as if budget support is not expanding and is less of a priority for bilateral as well as multilateral donors.

Before we leave programme modalities, two other forms should be noted. These are not of the same nature as the SWAps-GBS framework, but they are regarded as forms of aid which are programme-based and with high recipient government involvement, albeit with rather less focus on development activities. They have to do with debt and trade:

(1) *Debt relief*: This involves payments by donor governments to retire part of the debt of recipient governments. It does not lead directly to any development activity being funded, but it may relieve budget pressures on a government facing high debt servicing and repayment costs so that it can devote more of its revenue to operational matters. As a government-to-government transfer there may be agreements and conditions made with regard to public sector reform or particular economic or governance measures to be undertaken by the recipient government in exchange for debt relief.

(2) *Import support*: This modality was used in the past to ease balance of payments problems for recipients. It involved either the provision by donors of important imported goods, including raw materials or capital goods, or financial assistance to purchase such imports. This approach has not been used much in the past 20 years.

Programme decisions, learning and management

Programme support in its different forms involves an evolving relationship between donors and recipient states. It is an iterative process in that, as programmes take place, the capacity of recipients to manage aid funds hopefully is improved and, as a result, higher-level modalities can be put in place. In this sense, there is an intended transition towards GBS as the ultimate and desired mechanism for aid delivery. SWAps are seen to be an important intermediate step between projects and GBS because they involve a first significant stage of recipient-led aid delivery. In practice, however, the transition to GBS has been far from complete and SWAps have tended to be the most common higher-order programme modality.

In the process of building these higher-order modalities, there are important assessments and decisions to be made regarding the handing over of greater levels of responsibility to recipient agencies. Agencies such as the World Bank use certain processes to decide how aid should be delivered. There are some critical stages and requirements in these (Koeberle and Stavreski 2006). Firstly, it is important for the government to have in place agreed structure for 'general and sectoral policies and budget priorities' (or failing 'general' policies and priorities, there may be sectoral agreements in place). These amount to agreed Poverty Reduction Strategy Plans or national strategic development plans, which set out policy priorities and strategies and which donors have examined and endorsed. Without these in place, any higher-order modalities are not used: projects are used instead. Secondly, there is an assessment to be made regarding the overall macroeconomic environment and the robustness, capacity and record of economic management by the government, along with, critically, the assessment of 'public financial management systems'. These involve donors making a judgement regarding the ability of recipient public institutions to manage development processes. Inadequacies at this level can turn attention away from an overall government focus (potential GBS) to more sector-oriented strategies (SWAps) or, if there are doubts at sector level, again projects are the fall-back option. If, however, institutional capacity and transparency are approved of, then a final stage is to consider whether benefits are likely to result from a programme approach, in terms of increased local ownership, cost savings etc. If the answer is in the affirmative, then GBS or SWAps may be put in place; otherwise there is reversion to projects.

A key system that has become established in the process of assessing public financial management is the PEFA (Public Expenditure and Financial Accountability) mechanism. PEFA was established in 2001 by a partnership of seven donor agencies including the International Monetary Fund (IMF), World Bank and mainly European government agencies. It operates within the World Bank. It has developed a standard quantitative methodology for assessing the performance of financial management systems. 'It identifies 94 characteristics (dimensions) across 31 key components of public financial management (indicators) in 7 broad areas of activity (pillars)' (PEFA n.d.). PEFA assessments – and a rating score – have become critical for recipients. Assessments are repeated over time to give an indication of progress or not in improving financial management. Over 600 assessments have been conducted since 2005, mostly in the developing world, although Norway completed a self-assessment in 2008. Good scores (such as achieving an 'A' grade for key dimensions and showing improvements over time) help convince donors that a recipient government is capable of managing the resources it has and aid it receives. Good scores support the transition to higher-order aid modalities. Commentators have noted that PEFA has become a system that focuses on meeting an external idealised blueprint that does not fit all contexts well and undermines local ownership (Hadley and Miller 2016).

A point to be noted from this decision framework is that, should inadequacies be noted, donors do not simply revert to projects and give up on recipient governments. Instead we can see the importance of capacity-building measures. These aim to increase the capacity (the number of staff and quality of equipment) and capability (the skills of staff and quality of leadership) of government institutions. Training becomes crucial. Donors will support training through measures such as scholarships for tertiary study abroad or in-country programmes to teach particular skills and approaches. To a large extent this capacity building amounts to an alignment of recipient staff and institutions to recipient systems and ways of working.[2] Another feature of this framework is that much of the progression through the process is predicated on trust: can donors trust local systems and officials to work effectively, transparently and within acceptable levels of risk. In particular, they ask questions regarding 'fiduciary risk': are the aid funds likely to find their target and are they likely to achieve their desired objectives?

The shift away from programmes and the rise of retroliberalism

Behind these programme modalities and decision processes, which were being pursued in the first decade of the 2000s, lay the Paris Declaration of 2005. As we saw in Chapter 4, this agreement was brokered by the DAC and aimed to improve aid effectiveness. Its key principles – ownership, alignment and harmonisation, in particular – took overt control of the aid process away from donors and instead stressed the importance of recipient-'owned' development. Donors would respect and work with recipient institutions and systems (as long as they met recipient standards). The targets for achieving these principles also reinforced the transition to higher-level modalities. They involved goals such as the move away from tied aid and the use of local procurement systems. The Paris Declaration thus set the foundation and principles for the move to programme modalities and, ideally, GBS.

High-level programme modalities, backed by agreements, such as the Paris Declaration and its subsequent iterations at Accra and Busan, seemed to involve a remarkable degree of international consensus regarding what was best practice for aid delivery. It seemed to set an optimistic and forward-looking course for aid, building recipients' ability to set their development agendas, tackle poverty alleviation on a large scale and, eventually, take over the whole process without the need for donor support.

However, as we have seen, the global financial crisis of 2007–08 and political changes that followed undermined this consensus and reversed many of the positive reforms that had taken place. It also brought to the surface some of the doubts that had existed with regard to programme modalities and the neostructural framework. Firstly, donors realised that programme modalities and recipient ownership (together with alignment donor harmonisation) led to a diminution of their ability to exercise influence and control over their aid budgets. Aid had been an important foreign policy tool, used to garner political support and leverage on the global stage or improve economic prospects through trade and tied aid. Instead, with these high-level modalities, donor aid budgets were locked into long-term commitments with little room to manoeuvre, further negotiate or steer them so that benefits could flow back to donor economies.

There were also disadvantages for others. The central relationship with programme modalities was between donor agencies (bilateral and multilateral) and recipient governments. This left less space for civil society. Development NGOs had not been included in the Paris accord in 2005 and had to lobby hard for a recognition of the role of civil society at Accra and to be heard at Busan in 2011. Similarly, though often coming from a quite different approach to aid and development, was the virtual absence of the private sector, particularly in recipient countries. This sector, as with NGOs, had flourished with neoliberal reforms in the 1990s, but felt rather locked out in the neostructural environment. Some donor governments wanted a much more direct role for the private sector – something they pushed hard for in Busan – but this of course meant a weakening of the state-oriented programme modalities being put in place. Finally, there was a concern on the part of OECD donors that other donors, especially China, were not part of the global consensus on aid and that they should be invited to join. At Busan, China observed but decided not to become a signed-up member of the DAC-led approach to aid. The consensus was not complete and, with it fraying at the edges, donors seemed to lose enthusiasm for the Paris-inspired move to long-term, high-level, programme-based ways of delivering aid. In essence it is this set of circumstances that paved the way for the retroliberal regime as we discussed in Chapter 4.

Public-private modalities

Programme modalities did not end with the global financial crisis and the rise of the retroliberal era in aid – in fact most SWAp and GBS commitments continued and some expanded – but as they finished their funding cycles, donors appeared less willing to pit the bulk of their aid resources into these state-centred forms of aid delivery. Instead, they turned to new ways of thinking about how the private sector could be involved in various ways to augment aid activities. This has drawn from wider moves to promote public-private partnerships (PPPs) and there has been a trend to promote these as a way to finance development projects, particularly in the infrastructure sector (Romero 2015; Schur 2016).

The trend towards more public-private collaboration in aid funding has occurred alongside the rise of new discourses and methodologies for aid management. One example is the use of 'value for money' (VfM) approaches. These put much emphasis on material and readily quantifiable activities and on stringent scrutiny of costs (DFID

2011a; MFAT 2011). We have also seen a discursive shift to the use of 'results-based' narratives, again reinforcing the focus on activities which involve more short-term, physical and measurable outputs. Under these rubrics for aid funding and management, there appears to be less room for development strategies which focus on long-term qualitative changes, such as attitudes to girls' education, gender-based violence or acceptance of the ecological value of forests.

New aid modalities have begun to appear though many, as yet, remain in their infancy and few have become firmly established. Currently, the aid environment is characterised by a continuation of the state-centred programme modalities alongside these new and emergent ways of delivering aid and involving the private sector much more explicitly. In this section we examine firstly the international discussions regarding new forms of raising finance for development before we explore four examples of the way private sector partnerships are evolving.

Unlike the neostructural period and the focus on programme-based aid, the retroliberal period has not had the same level of open global discussion and agreement. There is no equivalent to the Paris Declaration of 2005 to serve as a set of guiding principles to which most donors and recipients are signatories. However, in 2015, an international meeting was held in Addis Ababa which explored the issue of financing for development (United Nations 2015a). This meeting did not result in a clear set of approaches or binding recommendations, but it did open the door to a wide range of funding options in the future. It turned attention away from simple state-state transfers and instead called for more diverse forms of funding involving donor agencies, recipient governments and the private sector (both local and global). The Addis Ababa meeting, then, is notable not for what it achieved in terms of clear guidelines but more in the way it marked a shift in thinking towards wider and more fluid partnerships between the public and private sectors. We now turn to some of these emerging new modalities for aid financing and delivery.

Private contractors

Private contractors have long been used by government aid agencies to deliver aid. They help reduce overhead costs, especially the cost of permanent head office staff, by shifting them to fixed-term contracts, and they limit the extent of government direct involvement (and, some might argue, responsibility for failures) in projects. Contractors are not bound by the same rules as government agencies (for example, they may pay their chief executives well above government salaries!) and they may be more nimble as a result. Although private contractors have been used for many years, it appears as if their use has been increasing, particularly by USA and UK (Roberts 2014).

USA is the world's largest ODA donor and by far the largest user of private contractors (Roberts 2014). About a quarter of the US aid budget in 2016 was channelled through for-profit firms and this share has risen appreciably since 2008 (*The Economist* 2017). For USAID most contracts are tendered on the basis of 'indefinite quality contracts' (IQCs) which provide multi-year funding typically to large contracting firms, who may, in turn, use subcontractors on IQCs (Villarino 2011). In 2016 some $US4.68 billion of aid went to such contract funding and, of the top-20 contractors, only one (Kenya Medical Supplies Authority) was based outside USA. The largest contractor, Chemonics International, received over $US 1 billion in 2016 and works in 70 countries on a wide variety of projects and sectors (Orlina 2017). Along with other large for-profit companies, such as Tetra Tech Inc, DAI, Abt Associates and AECOM, Chemonics forms a very large network of US companies that depend on the American aid budget (Table 5.1). In the UK, where 22 per cent of UK's bilateral spending went to private contractors in 2015–16, the use of contractors has risen sharply in recent years and some ten leading companies account for half of all contracts (*The Economist* 2017). It is also noticeable in Australia where firms such as Cardno are used increasingly by the government.

Aid contracting can be seen as a means for aid funders to ensure that their budgets are spent efficiently at arm's length by contracted organisations. They reduce the need to maintain large government aid agencies and they are manageable in terms of accountability and fixed commitments through the mechanism of legal contracts for delivery. Furthermore, the established consulting firms are well known to USAID and have close and trusted working relationships. Yet the contracting arrangement also helps channel very large shares of the international aid budget back to the donor's own economy. Contractors are usually involved in providing technical expertise or training (rather than the expensive physical hardware of development) so, although they provide advice and management

Table 5.1 ***Top ten aid contractors in USA 2016***

Name	*Location of Headquarters*	*USAID Funding in 2016*
Chemonics International, Inc.	Washington, D.C.	$1,009,133,442
Tetra Tech, Inc.	Pasadena, California	$471,061,443
DAI	Bethesda, Maryland	$343,817,396
Abt Associates	Cambridge, Massachusetts	$154,737,381
AECOM	Los Angeles, California	$148,843,815
Creative Associates International	Washington, D.C.	$146,245,498
Partnership for Supply Chain Management	Arlington, Virginia	$136,215,632
Kenya Medical Supplies Authority	Nairobi, Kenya	$122,652,321
FHI 360	Durham, North Carolina	$100,463,745
RTI International	Research Triangle Park, North Carolina	$95,373,963

Source: Orlina (2017)

services for aid deliveries to recipient countries, their salaries and fixed costs are spent in Washington DC, Maryland, Massachusetts or California (Roberts 2014). Although contracts are let through a competitive tendering process, in practice it is large Western firms who are best able to file the complex and expensive bids, and thus, using these contractors 'ties' development aid back to the donor economy.

Furthermore, the nature of contracts and the reliance on technical assistance programmes, means that aid delivered in this way marks a trend back towards project modalities: fixed-term, tightly managed by the donor (or donor contractor), a focus on pre-determined outputs and with limited engagement with, or attempt to build the capacity of, local institutions.

Public-private consortia

Another approach by aid agencies with regard to involving the private sector has been to act as a 'broker' in building partnerships in aid projects. Donor governments may still provide the bulk of the funding but, rather than simply disburse the funds to recipient governments, they attempt to bring together different parties to form consortia, each bringing different advantages and functions. Thus, for example, they may seek to involve a development NGO because of their ability to liaise with local communities, they may bring on board a local government department to help ensure public services are aligned, and they will want to have private companies involved often as the (commercial) providers of new services or hardware. In a way, this approach harks back to the old approach of integrated rural development: donors help build and co-ordinate a variety of institutions and projects to work together around a central activity.

One example of this approach is the UK's 'Invest Africa' programme (DFID 2017). This foresees FDI as the leading instrument for development: 'Invest Africa aims to address the barriers to significantly increase Foreign Direct Investment (FDI) in manufacturing sectors in Africa in order to help drive the creation of more, better and inclusive jobs in key focus countries' (ibid: 2). The UK aid programme, through DFID supports the programme in four main ways:

- '*Investor Engagement:* promoting awareness of opportunities, matching investors, buyers and countries.
- *Technical Assistance*: flexible technical assistance ... to strengthen policies, institutions and capacities to attract and retain increased FDI in manufacturing.
- *Political Engagement and Policy Dialogue*: Policy dialogue, raising awareness of potential investments and the binding constraints to realising investments, enhancing political commitment to economic transformation through FDI.
- *Coordination, influencing and knowledge management*: Promoting the overall objectives of Invest Africa, while crowding in and leveraging in additional funds, influencing strategic partners and programmes...' (DFID 2017: 3).

This example shows how aid is being used actively to help promote and augment private investment (foreign and domestic) and help ensure that it has support from recipient states. The resources of the aid budget and the discourses of development ('the economic transformation needed in Africa to create more and better jobs for the future and set countries on a trajectory out of poverty' – ibid: 3–4) are openly being used to support and subsidise the commercial interests of British companies investing in Africa.

Public-private consortia involve a mixing of different discourses of development. They may appeal to conventional notions of well-being and poverty alleviation (improving literacy, access to health services, roads or electricity), but this is usually alongside an articulation of the centrality of economic growth in the development process. Similarly, there is much more rhetorical use of the way small- and medium-sized enterprises, and the involvement of more people in market transactions, help build economies. This seems to emphasise the need to support and expand the private sector within recipient countries. Yet, as with the use of private contractors, the reality is of using donor-based larger private companies as the key agents of economic growth.

These consortia can also be seen as an expansion of the private contracting approach. Here, rather than just disbursing aid funds to private companies to deliver aid projects, donor agencies do more than this. They can undertake the feasibility studies and meet other early stage costs that prepare the ground for later private sector involvement. They also build a set of relationships that support the operation of private companies. The experience of the donor government agency, where aid is seen as part of a broader mission to promote national interests through trade and diplomacy, is used to bring together different parties in a set of joint activities. Public agencies – and NGOs for that matter – are often better suited to this sort of relationship building because of their long-standing involvement and relationships in developing countries, whereas private firms do not usually have the same level of experience or networks into communities, government agencies or even local businesses. So, consortia can be seen as donor support for the private sector through aid project funding plus relationship brokering (Box 5.2). These can be of significant benefit to foreign companies seeking to expand their operations in the Global South.

Box 5.2 Development consortia: Fonterra and New Zealand aid

New Zealand has a development agency inside its Ministry of Foreign Affairs and Trade (MFAT) and the Ministry's overall mission is to act 'in the world to make New Zealanders safer and more prosperous'. The country is also home to one of the world's largest dairy companies, Fonterra. Fonterra began as a co-operative of dairy producers within New Zealand, but has expanded with investments overseas and its growth strategy is focused on processing and supply chain management rather than an emphasis on farm-level production, particularly in the developing world. Following a memorandum of understanding between MFAT and Fonterra in 2014, there has been an expansion of the country's aid programme into new countries (Sri Lanka and Ethiopia – where Fonterra is planning new operations) and the inclusion of dairy development in its aid agreements with Philippines, Indonesia and Fiji.

MFAT has called for and let tenders for consultants to advise on dairy development in some of these places and it has also encouraged some NGOs (to which it provides funding) to become more involved in agricultural work. The aid programme, then, does not directly support Fonterra's commercial operations but, in using narratives of small farmer development or supply chain improvements, it has invested in activities which align in-country producers with Fonterra's plans to expand processing plants.

Across both Fonterra and MFAT we can see new framings of development and aid. Both might talk about the health benefits of poor people getting access to better supplies of protein through increased dairy production[3] (something that NGOs and local governments can promote) or the involvement of more small farmers in increased and better quality dairy herds (through aid-funded technical assistance and development of the sector). However, there appears to be an underlying and largely disguised factor which is related to the building of milk processing plants overseas by Fonterra and their need both to build and maintain milk supply chains to their plants and build local demand for their products. Furthermore, with concentration on increasing production and linking farmers to new markets, it seems as if some important elements of development, such as a concern for the social, gender and environmental dimensions of change, have been overlooked (Edwards 2019).

The use of consortia, then, represents a new and innovative approach to donor aid funding. The role of aid agencies is to provide much of the basic funding, but then use its networks with NGOs, recipient governments and its own private sector to broker large-scale agreements that typically involve a range of projects. Much of the funding ends benefiting donor country companies – and may help promote their efforts to expand their operations in developing countries – but it is done in the name of bringing development benefits to the poor. Again, we see a reversion to project modalities and a relative lessening, but not elimination, of the role of recipient governments. Private sector growth is the main priority and whilst this may involve some local enterprises, donors are keen to back companies from their own countries to do business overseas.

Such ways of operating, as these consortia do, also changes the way aid and development are portrayed. Private companies wanting to do business and make profits do so when economies are expanding and where local institutions (and things such as law and order and protection of property rights) are secure. They do not work so well when there is extreme poverty (people do not have money to spend) or governance is weak. So, although there is some use of poverty narratives to justify aid spending, the rhetoric of development has tended to shift in subtle ways to emphasise growth and prosperity in local economies and support for expanding local economies rather than alleviating extreme poverty.

'Blended finance': development impact bonds

A third, and as yet rather untested, mechanism for building PPPs for aid delivery involves the use of private investment to underwrite development interventions. It is 'blended finance' because it sees private capital (from investors or philanthropists) being used to fund an activity, whilst donor funds (either private or public) are only paid out to repay and reward the investors if the project succeeds. Development Impact Bonds (DIBs) draw on the model of social impact bonds in places such as USA and UK where several parties are drawn into an intervention to secure a defined social goal (Fraser *et al.* 2018, Berndt and Wirth 2018). They are 'another example of the development community's pivot toward results-based financing and greater private sector engagement' (Glassman and Oroxom 2017). At the macro level, a public agency will reward an agency (an NGO or a government department) if it oversees a project that brings about positive change as indicated by certain pre-defined measures. This might be an increase in the number of girls attending school or a lowering of prisoner recidivist rates, for example. This is termed a 'pay for success' approach with the donor agency only paying for the intervention if it succeeds in meeting its targets. In theory, the provider is incentivised to be flexible and innovative in order to achieve the results (rather than focus on activities and managing inputs). The interim funding for the implementing agency however is provided by bonds taken out by private investors, usually through the intermediary of a financial company or bank. These investors will only receive a repayment of their investment, and a healthy premium, if the targets are met. If they are not, they lose their money and the donor does not pay up. In effect, they are 'investing' in a gamble, putting faith in the ability of the implementing agency to do their job. Investors may be motivated by social concern – feeling good about supporting a needed form of social development – or they may simply see a good opportunity for profit.

The social impact bond model has been extended to DIB though these are not yet common. In these we have the donors (they can be a donor agency – government or NGO), the implementing agency (a local NGO or a local government agency), the investors (usually foreign) and the financial intermediaries (companies which design the schemes, manage the bonds and monitor the performance – all for a set fee). Donors only pay if the development interventions are successful, implementing agencies have a strong incentive to achieve success so they secure future contracts, investors will gain a profit if the project succeeds, and the financial intermediary will draw its commission whatever happens. One of the first DIB schemes was 'Educate Girls' in Rajasthan, India. This was launched in 2015 and aimed to improve education (in terms of enrolment and retention) for 15,000 pupils. It involved an investor to pay the upfront costs, a service provider, an outcome payer (the Children Investment Fund Foundation) and an independent evaluator. After two years, the scheme appeared to be well on the way to meeting its targets so the investor would recoup their funds and the outcome payer would pay for a successful outcome (Loraque 2018). Gradually we are seeing further examples of DIBs being established (Box 5.3) though they are not yet large in terms of total volumes of aid.

DIBs can be seen as a financial product. They attract investors to finance development activities – thus this is seen to broaden the base of financing for development. Yet, in practice, all they are funding are the working costs of the project and bearing the risk. Donors still pay for the development activity but are now in a way paying more both to cover the premium to investors (in recognition of the risk being borne by them) and the commission of the intermediary. Furthermore, implementing agencies continue to do their jobs, now just reporting to a new paymaster (often with strict conditions and targets). Another implication of DIBs is that they take aid modalities back to project-based activities with defined time frames and quantifiable outputs. They eschew the longer-term programmes of change that may involve more qualitative changes (such as advocacy or attitudinal transformations) and they do not involve wider recipient government involvement as programmes do: governments are put in the role of contracted service providers, not 'leaders' of development strategies and nation-wide changes.

Box 5.3 Development impact bonds: the Cameroon Cataract Bond

The Cameroon Cataract Bond is part of a larger set of activities to tackle the high prevalence of vision impairment in Sub-Saharan Africa by organisations such as the Africa Eye Foundation. This foundation established the Magrabi ICO Cameroon Eye Institute (MICEI) in Cameroon to provide training for ophthalmology staff and treatment for patients. The founder and main donor for the Magrabi hospital, Dr Akef El-Maghraby is an ophthalmologist stated:

> *My goal is to make high-quality eye care accessible to the people of Cameroon regardless of their ability to pay. We are grateful to our stakeholders for supporting the hospital's path to sustainability so that we can continue to deliver on our promise. MICEI … lays the foundation for a network of integrated and self-sustainable eye hospitals the organisation plans to build across Africa.*
>
> (*African Review* 2019)

With the hospital in place, there was a need to attract further funding to provide training for surgeons, nurses and allied staff and to support the costs of surgery and care. The DIB mechanism was launched in 2018 in the form of the Cameroon Cataract Bond, a 'pay-for-performance' loan. It brings together investors, donors and eye care specialists. The Conrad N. Hilton Foundation is providing most of the outcome funding pool (along with the Fred Hollows Foundation and Sightsavers) and will pay out if the targets are reached. Costs of the project are being met in part by a $US 2 million loan from Overseas Private Investment Corporation (OPIC), a US government agency (OPIC 2017). MICEI is the service provider and is supported by private and public investors as well as the OPIC loan. 'The project aims to provide 40 per cent of all surgeries for the people in the poorest two groups of the national population' (*African Review* 2019).

In its first year under the scheme MICIE screened over 50,000 people and completed more than 2,300 cataract surgeries. Low-income patients receive their operations either for free or at a subsidised rate. Four weeks after surgery some 92 per cent had achieved a 'good visual outcome' and the quality of surgeries exceeded WHO standards for such operations.

This DIB scheme, still in its early days is an example of how an active organisation has sought to find new ways of attracting funding (Glassman and Oroxom 2017). Much of its support has come from traditional forms of donations from various individuals and charities and it has had government support also. But it has seen in the DIB mechanism a way for its core work to be funded by overseas benefactors – the Conrad N. Hilton Foundation – on the basis of results achieved. It can present to donors a narrative of results achieved, rather than inputs needed. It has also been able to combine the DIB income stream with more conventional ones, such as (subsidised) user-pays and private donations.

Viability gap funding

Finally, we can identify a fourth public-private aid modality that is emerging in recent years. This sees donor governments directly subsidising the operations of private companies through what has become known as 'viability gap funding' (VGF). This approach, which seems to have developed particularly in India in relation to PPPs for infrastructure development, rests on the assumption that private companies should fund certain development activities (such as the construction of roads) that will have distinct development benefits, but their decision to invest may be negatively swayed by an assessment that their profit margins may not be sufficient nor secure enough. There is, for them, a gap in the viability of their investment. This gap can be filled by a sympathetic donor who can see that a subsidy for the private company would make the project viable. Projects that are economically – or developmentally – justified, but not financially viable are made viable by government assistance (Schur 2016).

This approach represents perhaps the highest level yet of direct use of aid funds to support the private sector. In this, donors regard private companies as key drivers of development, providing goods and services, generating employment and involved in major development projects, such as infrastructure (Box 5.4). But they also recognise that private companies face risks and uncertainties. If these can be reduced, then viability gaps can be closed and investment secured in circumstances where, in the past, it might have decided not to proceed. In short, aid is used to bolster the profitability of private investment so that economic benefits (employment, growth, better services) can trickle down to the local population. It represents a significant shift in the locus of development: rather than committing to partnerships with recipient government agencies, through SWAps and GBS, to large-scale and long-term development programmes, VGF has moved aid relations to the investment decisions of private companies seeking profit in developing countries, helped out by their supportive governments.

An example of this type of approach, even if not directly identified as VGF, is the OPIC, 'a self-sustaining U.S. Government agency that helps American businesses invest in emerging markets' (OPIC 2017). OPIC has been operating for nearly 40 years and it 'provides

Box 5.4 DFID and the private sector

The British aid programme has been prominent in the past decade, increasing its budget substantially and achieving the 0.7 GNI target. These increases have been accompanied by a 'retroliberal' shift towards greater support for the private sector and a more explicit alignment with the UK's wider economic interests. There have been several instances of how this has been manifested with the country's aid budget being used to support UK businesses, such as supermarket chains, working with their suppliers in the Global South to improve standards (through the Trade in Global Value Chains Initiative). There has also been engagement with the London Stock Exchange to train financial sector professionals in Tanzania both to assist with their own stock market development, but also to develop longer-term relationships (Mawdsley *et al.* 2018, also Mawdsley 2015).

The most controversial element of this support for private sector development though has been the use of substantial aid funds to increase the volume of investments by the former Commonwealth Development Corporation (now the CDC). The CDC, operating within the UK aid agency (DFID), has a mandate to invest in private companies (both British and overseas) with the aim of reducing poverty. This aim is seen as being fulfilled when companies grow and create employment. In 2015, it announced that CDC would receive an additional £735 million in funding over three years through the aid budget. At the time, Department for International Development (DFID) justified the funding by stating: 'This will ensure countries can grow and trade their way out of poverty while building future markets that British businesses can compete in' (Anderson 2015a). They also pointed to the claim that their investments employed more than a million people, especially in Africa and South Asia.

The investments of CDC, however, have received criticism. They are often made without specific reference to poverty alleviation. Indeed, they tend to target business opportunities with a good chance of success and return. Investments have included ten gated housing communities in El Salvador, a large-scale and upmarket housing and retail development in Kenya, and beachfront housing development in Mauritius (Provost 2014).

Whilst such private sector developments could be seen to generate employment in the construction sector, they would hardly seem to be focused on poverty alleviation and their inclusion in the ODA accounts seems questionable. One critic believed the UK government had exported a 'highly financialised, highly unequal, highly ideological form of "development" which helps big business, not ordinary people' (Nick Dearden, director of the World Development Movement, quoted in Provost 2014).

The Independent Commission for Aid Impact (ICAI) assessed DfID's more general engagement with the private sector in 2015 and gave it a rating of 'amber-red' (the second-worst category). One journalist commented: 'Part of the problem is that DfID's poverty reduction targets are out of sync with the profit-making imperative fundamental to private business' (Anderson 2015b).

businesses with the tools to manage the risks associated with FDI, fosters economic development in emerging market countries, and advances U.S. foreign policy and national security priorities' (ibid). It charges fees and operates as a lender of capital (as in the Cameroon Cataract DIB – see Box 5.3) and thus raises its own revenue, not drawing on government aid funding. It does, however, fit well with these new modalities, showing a willingness of donor states to support the operations of their own private sector to do business in the developing world in the name of both 'development' and (donor) self-interest.

A more explicit use of VGF is by the Private Infrastructure Development Group (PIDG) 'a multi-donor organization with members from seven countries (Australia, Germany, the Netherlands, Norway, Sweden, Switzerland, United Kingdom) and the International Finance Corporation of the World Bank Group' (IISD n.d.). PIDF will grant up to $US 3 million in VGF per project. 'It targets pro-poor infrastructure projects that are economically viable in the long term but require initial funding for commercial viability and acquisition of private sector investments' (ibid).

The VGF approach might be understandable in efficiency terms if there was a transparent and competitive process for funding, as there seems to be in the PIDF process. Donors would fund whichever company on the open market could prove they were most likely to meet the development objectives of their investment and which had the smallest viability gap to fill. In some cases, this may be partly the case, for example when the Government of India will assist local companies, following a competitive bidding process, to undertake infrastructure projects. However, we are beginning to see the approach being used by Western donors in relation to support for companies from their own countries to do business overseas (such as, in part, the UK's Invest Africa programme – above). It seems that if a narrative can be constructed that such private investment will bring development benefits (such as employment generation or better services), but it is not yet financially viable, the donor VGF can be used to tip the balance in favour of the investment. Aid funds are used to subsidise donor country private investment and enterprise in developing countries.

Conclusion

In this chapter we have examined different modes of delivery of aid from projects to programmes to PPPs. Until about 2008, it was tempting to see a progression in modalities, a realisation that a steady and considered transition from projects to programmes (SWAps then GBS) represented an improvement in aid and development effectiveness. Best practice and the lessons of past failures seemed to point to the need to work with local institutions and let them take the lead and 'own' the development strategies that aid supported. Also, it was recognised that aid needed to 'scale up', to adopt longer time frames and consider long-term development outcomes, rather than focus solely on producing concrete and defined outputs over a limited project cycle. Aid should support country-wide programmes of change to have broad impact and really tackle macro-scale issues such as poverty alleviation, not just worthy local projects that had limited impact nationally.

However, this assumption of progress in aid modalities and practices has been overturned by changes to modes of delivery in the past ten years. The desire of many donors to retake control over the process of aid delivery (rather than leave it recipient governments) and use it to promote wider national interests, including the support for their own private sector, has seen a return to lower order modalities and much more complex relationships amongst governments, civil society and for-profit businesses. The conversation regarding aid delivery has changed markedly and there has been a shift away from global-level agreements that provide direction for these relations, as was the case in the 2000s following the Paris Declaration.

These new forms of PPPs in relation to aid are characteristic of the retroliberal approach to aid. They see aid donor governments using the aid budget to promote the interests of their own private sectors in the name of 'doing development' and assisting developing countries. They play very much into the retroliberal narrative of 'shared prosperity' and are a part of the trend towards financialisation of aid. As we move along the scale from the use of private contractors, to consortia and impact bonds to VGF, we see an increasingly explicit subsidy and support for the private sector. There is a move to more 'supply-led' aid; aid that is constructed on the basis of what donors (and their private sector) can and want to achieve in partner countries, rather than in response to the requests from those partners. Furthermore, as these private-public mechanisms for aid continue to evolve, we see a further retreat from programme modalities back to projects as the main vehicle for aid delivery.

Therefore, rather than assume that aid delivery will or should simply follow the signs towards better aid practice and more effective development through the use of higher-order modalities, we should accept that the different modes open to donors will all be used in different ways and at different times, depending on the current political and ideological climate. Furthermore, there has been no simple replacement of one dominant modality by another as we shift from one aid regime to another. During the neostructural approaches of the early 2000s it seemed as if there would be a move up the modality scale as projects gave way eventually to budget support. However, this has not happened and has been partially reversed. Modalities overlap in time and space. Budget support remains a very important mechanism used by several donors and is favoured by most recipients; similarly, projects remain as the staple mode of operation for many agencies, especially in civil society or where there is not a strong relationship of trust between donor and recipient. What has happened in the past decade, however, is the emergence of new modalities – not yet dominant by any means – that may signal a new direction for aid operations; modalities which seek to combine public and private finance much more than in the past.

Summary

- Modalities are set of modes of practice that are utilised in the delivery and application of aid funds.
- There exists a range of modalities that vary from lower-order to higher-order approaches. At the extreme of this continuum are projects and general budget support, respectively.
- There are a number of modalities that exist in between these two extremes and combine elements of projects and programmes.
- Lower-order modalities, centred on projects, bring both costs and benefits. They are discrete, have specific objectives and an end date. On the other hand, they are not long-term enough to tackle the deep-seated roots of poverty and underdevelopment and often by-pass governments.
- Higher-order modalities are longer-term. They include SWAps, GBS, as well as debt relief. The former two became common

in response to the rise of the MDGs targets and the Paris Declaration and focus on aid ownership and effectiveness.
- SWAps involve sectoral approaches where control is handed to recipient governments over time. GBS is considered the optimal form of aid where governments have the most control over budgets and their disbursement.
- Humanitarian relief accounts for approximately 13 per cent of ODA at the global scale. Humanitarian aid arises when there is a disaster or an emergency, and essentially become project-focused in terms of their methodology.
- Such disasters are often not 'natural' – although such things may precipitate an emergency. Humanitarian crises often result from more profound underlying causes – such as conflict, inequality, non-sustainability – all of which have political causality.
- There has been a shift recently in terms of the most common modalities. Neostructural approaches favoured programmes and longer-term approaches. The retroliberal shift has seen the rise of new project approaches based on private-sector investments.
- Public-private modalities have accompanied the rise of retroliberal aid. These include private contractors, public-private consortia, blended finance, and VGF. These place more control in private hands and have been accompanied with a shift to narratives involving 'prosperity' as opposed to poverty reduction.

Discussion questions

- Define an aid modality and give examples of lower- and higher-order modalities.
- What are the costs and benefits of project modalities?
- What are the costs and benefits of programme modalities?
- To what extent are humanitarian disaster natural disasters? What role do politics and deeper development challenges play?
- Discuss and explain the difference between modalities applied in the neostructural and the retroliberal aid regimes.
- How has the narrative accompanying aid shifted with the move to public-private modalities and how can this be criticised?

Websites

- PEFA: pefa.org/about

Notes

1 The terminology used in the project management field varies. Many use the four-step initiation/planning/execution/closure framework. Here we also employ the identification/design/implementation/evaluation terminology, often used in development work. We also add the feedback loop through monitoring and evaluation to portray the idea of a cycle where lessons learned are applied to later projects. The addition of the outcomes/impacts element is to illustrate how the longer-term effects of a project are critical, but usually not well incorporated in the project cycle approach.
2 This is rather a mirror image of the Paris principle of 'alignment' which calls for donors to align with recipient government systems and strategies. In practice, capacity building sees local institutions fall into line with donor systems.
3 This story rather glosses over the dubious and contested promotion of the health benefits of providing infant milk formula – milk powder is a leading export product of Fonterra and much is turned into infant milk formula.

Further reading

European Commission (2018) *Budget Support: Trends and Results 2018*. Directorate-General, International Cooperation and Development, European Commission, Luxembourg.

Koeberle, S. and Stavreski, Z. (2006) 'Budget support: Concepts and issues', in Koeberle, S., Stavreski, Z. and Walliser, J. (ed), *Budget Support as More Effective Aid*. World Bank, Washington DC, pp. 3–27.

Reinikka, R. (2008) 'Donors and service delivery', in W. Easterly (ed.), *Reinventing Foreign Aid*. MIT Press, London and Cambridge, MA, pp. 179–199.

Roberts, S.M. (2014) 'Development capital: USAID and the rise of development contractors', *Annals Association of American Geographers* 104(5), 1030–1051.

6 Does aid work?

Learning objectives

This chapter will help readers to:

- Understand competing perspectives concerning the positive and negative impacts of aid
- Appreciate the arguments concerning the economic impacts of aid, and its costs and benefits
- Debate the impacts of aid on the notion of good governance, understanding both success and failures in this ambit
- Understand and debate the relationship between aid and poverty reduction
- Assess the evolving role of non-state actors, civil society and the private sector, in aid flows
- Consider the transparency of aid allocation within donor countries and whether this needs to be enhanced

Introduction

Given the huge volumes of aid that have flowed across the world for the past 50 years and more, we might well ask whether aid actually works. This is a question that has attracted much attention over the years from a wide variety of theoretical and political perspectives and the debate continues (Cassen 1994; Dollar and Pritchett 1998; Riddell 2007; International Poverty Centre 2007). Evidence can be applied to support different conclusions from different viewpoints. To an extent it depends what questions are asked, how they are constructed and – of course – who is answering them! However, here we will consider some key questions, and seek to be objective in answering them. There are in fact two sets of problems we must consider when answering the overall question 'does aid work?' The first set relates to the concept of aid and how it has been operationalised itself – is it problematic conceptually, or even philosophically and should it therefore be ceased? Do any observed problems have more to do with the way it has been applied and therefore is it possible to reform it? A second set of questions refers to the costs and benefits: if we accept that there are both positive and negative aspects to aid, how do we weigh these up? To answer these two sets of problems we separate the effects of aid into economic, political/governance, social impacts. We start by discussing how aid relates to economic growth and performance, and then turn to its effects on governance. Following this, we assess the effects of aid on poverty alleviation. In doing so, we consider the role of civil society and the private sector, both of which have been seen in positive light over the most recent decades in terms of being both targets of and facilitators of aid projects. We conclude with some insights into how aid delivery works with a particular focus on the performance of donors.

Aid and economic growth

In recent years aid has been conceptualised by some donor countries as a way of promoting 'sustainable economic growth'. It is believed that aid can promote growth by supporting the private sector and when growth occurs development benefits will follow (or 'trickle-down' to use the often-utilised and over-simplified terminology), particularly through the generation of wage employment and the positive economic multiplier effects associated with that. The supposed ability of aid to encourage growth has been a characteristic of early neoclassical views within modernisation theory and, though opposed by more fundamental neoliberal views, it has resurfaced in more recent retroliberal aid regimes (see Chapter 4).

The arguments in favour of using aid to promote economic growth rest on a fairly basic premise that developing economies lack key resources, particularly capital and technology. A well-known economic concept, the Harrod-Domar growth model, suggests that the rate of growth of an economy rests on two critical features: the productivity of capital (how efficiently capital is used) and the savings level (savings provide investment that can be used to increase the capital stock). Given that the productivity of capital is hard to change in the shorter term, the key element for economic growth is capital. When savings rates are low, as they are in developing economies, then capital can come from external borrowing, foreign investment or aid. This then sees a basic line of argument that suggests aid can help provide capital which then fuels economic growth, which in turns brings jobs, economic multipliers, and broader development (Todaro and Smith 2015).

The role of aid in promoting growth can come in several forms. At one end of the spectrum, it can be capital investment, such as the building of infrastructure (roads, energy supplies etc.); at the other, it can be in providing seed capital for small-scale savings and investment (such as in micro-credit schemes). In between it might see aid resources used to integrate and support private sector activities in development projects, so that private sector profitability is enhanced by being able to tap development funds. A recent study by Dalgaard and Hansen (2017), noting the renewal of interest in funding large infrastructure projects, found that the average gross rate of return on aid investments 'is close to 20 percent' (p.1012).

Other economic models provide a justification for aid in further ways. The Solow-Swan growth model (Solow 1956), seen as a counter to Harrod-Domar, was a more complex model to show the components of growth but it stressed technological progress rather than capital accumulation. For aid, this meant that economic growth in developing countries could be accelerated by promoting the transfer of technology from more progressive economies (and this was certainly embodied in Truman's post-war speech). Again, this could be in different forms: from encouraging foreign investment (applying modern industrial techniques); to providing incentives for technological innovation; and expanding higher education and vocational training. Many of these could be assisted by aid particularly through training programmes and scholarships to study abroad.

These justifications for aid to promote economic growth, however, were countered by criticisms, which were rather less optimistic and which saw potential economic drawbacks. Firstly, there is the danger that capital introduced from external sources, such as aid, will crowd out scarce local sources. In conditions where the rate of savings is low (most people need all their incomes to survive from day to day and only a relatively few are able to save a surplus), higher interest rates are an incentive for this rate to grow. However, when there is an influx of external capital, interest rates may fall, inflation may rise and good investment opportunities may be restricted. People may shift from saving to more consumption.

Similarly, there can be price distortions with aid which can negatively affect local economic growth. For example, and perhaps counter-intuitively, food aid can lead to future food shortages. When food is scarce, perhaps as a result of, say, drought, then food prices rise. When this happens, local producers who do have a surplus are encouraged to sell and receive a good return, encouraging them to plant more for the next season. But when large volumes of food aid appear and swamp local markets, food prices fall and local producers, whether larger-scale farmers or households producing small surpluses, are less inclined to plant for the future, getting their grain from aid outlets and turning their attention to alternative cash crops. Food production continues to fall, even when harvests improve and food aid can become a more permanent fixture. Similarly, aid projects which require that machinery, fertilisers, computers etc. be purchased direct from donors, by-pass local suppliers and again may depress local prices and disincentivise production. Local businesses face what they see as unfair competition and may suffer as a result.

Aid can also promote dependence and inhibit local development. Aid projects bring in new resources and ideas. If these prove to be useful, there will be a continued demand for these new products or sources of technology. Local sources and ideas are seen as inferior and the new activities that come with projects become closely aligned to external inputs. Furthermore, the viability of activities beyond the project's notional completion date can be threatened if the external sources of supply and subsidy are cut off. This can involve, for example, dependence on continued aid funding to maintain the overhead costs of a project (administration, maintenance etc.) even if the day-to-day activities are viable.

Finally, the question of inappropriate technology is raised. Aid based on donor-supplied technology can have the effect of distorting an economy towards forms of technology that are out of line with local conditions. Technological development in industrialised Western countries has usually been developed against a backdrop of relative labour scarcity (and higher wages), capital abundance and a skill base in the population aligned with the technology used. By contrast, developing countries, as aid recipients, usually have the opposite conditions: cheap and abundant labour, scarce capital and populations skilled in customary knowledge and low-tech activities but not in the use and maintenance of complex equipment. The result is that aid provides capital equipment that appears to provide immediate benefits, but which cannot be easily maintained into the future. These longer-term costs are not usually accounted for in project planning and impose a burden on the domestic economy.

There is conflicting evidence for the positive or negative effects of aid on economic growth. Some critics (for example Hughes 2003) argue that aid has inhibited or even reversed economic growth and there appears to be no clear correlation between levels of aid and economic growth at least in the short term. However, an International Monetary Fund (IMF) working paper in 2009, looked at such correlations over a longer term and found that a positive relationship did exist, albeit with a long lag-time of up to several decades (Minoiu and Reddy 2009). This suggested that aid could promote economic growth, but only if directed at development-oriented activities and some donors (such as Scandinavian ones) were more effective than others. The authors concluded that effective expenditures to promote growth

> *may support investments in physical infrastructure, organizational development, and human capabilities, which bear fruit only over long periods … our findings help counter claims that aid is inherently ineffective and aid budgets should be reduced. On the contrary, an increase in aid and a change in its composition in favour of developmental aid are likely to create sizable returns in the long run*
>
> (Minoiu and Reddy 2009: 17)

Burnside and Dollar (2000) also found a positive relationship between aid and economic growth within a 'good policy environment'. Furthermore, some recent studies have further supported the view that 'aid promotes growth in a statistically significant manner' (Mekasha and Tarp 2019: 14; also Galiani *et al.* 2017; Morrissey 2001; Arndt *et al.* 2015).

Nonetheless, there does not appear to be any clear indication as to what specific types of aid promote growth more than others. The debate on the relationship between aid and economic growth is likely to continue. For neoliberals, market failure is largely caused by intervention itself, whereas for neostructuralists intervention must take place in order to correct for those failures. In some ways the whole debate comes down to how you define and measure market failure and negative externalities (such as low savings rates, low levels of technology, and environmental problems) and what you consider can be done to address them. Those on the right-wing believe the market allocates best and intervention creates more problems whilst those more to the left believe that intervention is required to address the problems. A further arm of the left argues that aid creates dependency and foreign domination. In other words, the answer to the question – does aid stimulate the economy? – depends on how you deconstruct the question.

However, some lessons can be drawn from this relationship. Under some circumstances aid can improve economic performance, but it can also inhibit or distort local economies. Aid, in itself, may not be the best mechanism to promote economic growth – many would argue that trade and investment are better. Aid should work with, and not crowd out, local resources, enterprise and knowledge. It can fill some gaps and augment local resources. Arguably, though, it is perhaps best directed at helping to build the long-term capabilities of local economies, through infrastructure, education and training and institutional strengthening. All these require long-term commitments and relationships and there is apparently no 'quick-fix' solutions whereby aid can stimulate deep-seated and sustainable economic growth and development. However, it seems that there is robust evidence that aid does promote economic development, mainly through the way it helps build physical infrastructure and human capital. Arndt *et al.* (2015: 15) conclude:

> *There is no evidence that aid is detrimental. Aid has contributed to economic growth by stimulating its proximate determinants – e.g., physical capital accumulation and improving human capital, particularly education and health.*

Aid and governance

Whereas the debate concerning aid and economic growth is predicated on the assumption that economic growth is the key means to promote the process of development, other approaches to aid focus on less economic parameters of development such as human well-being or institutional capacity. One of the main approaches in this sense grew out of the post-SAP (Structural Adjustment Program) years of the mid- to late 1990s and stressed the importance of 'good governance'. In large part, this still had many neoliberal features – it believed that previous aid approaches had created and expanded a largely inefficient and bloated public sector that inhibited market-led development – but it recognised that states still had important functions to perform and it blended in concerns for promoting democracy, human rights and public participation. Its theoretical inspiration also came from a branch of economics, this time in the form of the 'new institutional economics' of the 1990s, pioneered by Douglass North and others (Martens *et al.* 2002). This put a focus not so much on market mechanisms but on the way institutions functioned, in particular their rules and social and legal norms. Although new institutional economics was concerned with a broad range of institutions, including the private sector, the basic ideas regarding the way institutions function influenced and suited attempts to rebuild and recraft the state following earlier neoliberal 'rolling back' of the state.

Good governance promoted a particular justification for aid to do with the way aid could help recreate and align state systems so that governments could operate efficiently and transparently, both allowing the markets to operate effectively and with security, and promoting citizen involvement and confidence in state systems. It rests on the view that states have a leading role to play in development and this was a key foundation of the 'neostructural approach' of the early 2000s. It involved a 'rolling out' of the state – introducing new ways of operating and new forms of regulation that created the conditions for the market to function effectively. It is also allied to the role aid can have in promoting democracy, often through the use of aid conditionalities to pressure states to introduce democratic reforms. Gibson *et al.* (2015) concluded that aid, through technical assistance, has indeed contributed to democratisation in Africa.

The main approach of the institutional strategy is to seek to reform the public sector in two main ways. Firstly, there is capacity building and enhancing capability. Capacity has to do with increasing the number of public servants working in defined areas and capability has to do with building the skills and experience of those performing critical jobs. Education and training are crucial for both, ensuring that there is a good supply of well-educated applicants for all positions and providing specialist skills where needed. The second strand is to do with the policies, processes and procedures of institutions. These were intended as systemic improvements, reforming the way institutions function. These involved aspects such as the separation of funding and providing of services; improved transparency and accountability through better information systems, standardised reporting and public communication; greater public consultation; an end to political interference in operational decision-making; and the use of international standards and systems to manage human and financial resources. Again, this strand rested on education and training, though here with the added strategy of using international consultants to offer advice on how the systems should work. An added attraction of these approaches for donors was that it helped align recipient government systems with those of the donors, so that officials on both sides could understand common principles and systems and donors could more easily track how aid money was being managed and allocated through recipient agencies. Thus, aid for public sector reform aimed at all facets of the state, from law and order, the judiciary, the legislature, the full range of line ministries, and policy making as well as service delivery.

As well as these approaches to public sector reform, the use of aid to support government functions has been justified in terms of the use of higher order modalities (see Chapter 5). Simply put, it is recognised that to achieve significant long-term and large-scale improvements in the welfare of a country's population, it is necessary to commit to substantial and sustained support for core government functions to do with human welfare, particularly education and health. The significant increases in aid following the turn of the new millennium were largely channelled, following the Paris Declaration, through recipient government systems in the form of SWAps and budget support.

However, there were and continue to be critics who suggest that greater funding for government operations and the use of recipient government agencies, even if reformed, is not an effective use of aid and is open to misuse and distortion (e.g. Moyo 2010). There are several aspects to these cautions and critiques. Firstly, without very close scrutiny, there is the issue of fungibility. Fungibility refers to the ways state funds can be shifted to other uses when aid is used to cover some budget items. Thus, for example, Official Development Assistance (ODA) is not used to fund expansion of the military of a recipient but if it contributes substantially to, say, health and education budgets, then a government may feel that it has some freed up resources (otherwise spent building hospitals and schools) to spend on new equipment for the army or air force.

Box 6.1 Aid to Afghanistan

Afghanistan has been one of the major recipients of ODA since its invasion by US-led coalition forces in the wake of the 9/11 terrorist attacks in 2001. Soon after the invasion, aid inflows rose steadily: since 2005, net ODA has exceeded $US 3 billion per year (Figure 6.1) – though military spending amounted to about $36 billion a year early on. Much of the aid has been spent on humanitarian relief and technical co-operation, though a great deal as gone to the everyday functioning of government. In 2008 it was reported that 90 per cent of all public expenditure came from international assistance (Waldman 2008: 1).

Figure 6.1 ***ODA disbursements to Afghanistan 2000–17 ($US mill constant 2017)***

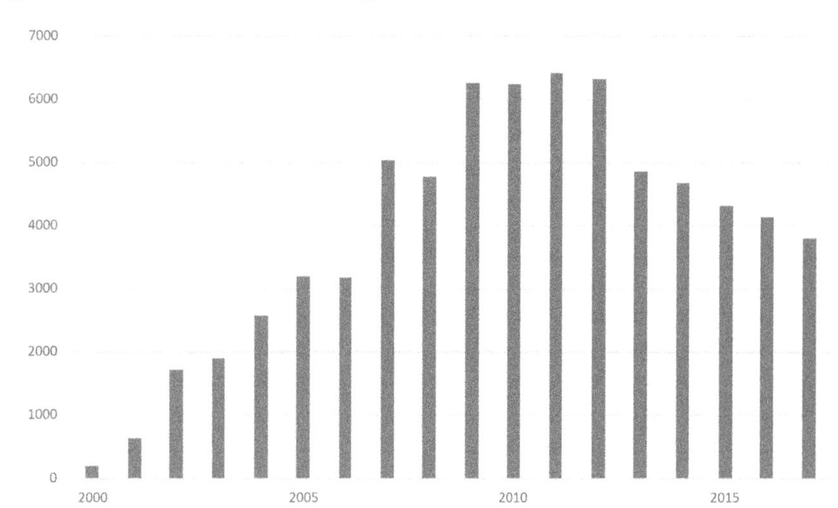

Source: www.stats.oecd.org.

A 2008 review of aid to Afghanistan in the early years concluded:

> *Much has been achieved in Afghanistan since 2001: there has been the establishment of democratic institutions and ministries, significant improvements in health care and immunization, the major expansion of primary education, the construction of roads and transport infrastructure, economic growth, and the formation of state security forces. There are many cases of well-delivered aid, for*

example in the education sector or in community-based rural development projects that are part of the National Solidarity Programme (NSP), which have made a significant difference to Afghan lives.

However, most Afghans still endure conditions of hardship and millions live in extreme poverty. Far too much aid has been prescriptive and driven by donor priorities – rather than responsive to evident Afghan needs and preferences. Too many projects are designed to deliver rapid, visible results, rather than to achieve sustainable poverty reduction or capacity-building objectives. One quarter of all aid to Afghanistan has been allocated to technical assistance – which is intended to build government capacity – yet much of such assistance has been wasteful, donor-driven and of limited impact. In the design or execution of projects, too often the promotion of the capabilities, status and rights of women is an afterthought or perfunctory consideration. Most aid has been directed to Kabul or other urban centres, rather than to rural areas where it is most needed and more than three-quarters of Afghans live.

(Waldman 2008: 2).

Later (Fayez 2012: 70) noted that 'Afghanistan is transforming into a black hole because billions of aid dollars are being spent without any remarkable positive effect on the standard of living of the people and the stability of the government'. Yet, the heavy expenditures have had positive effects in some areas: 'A lot has been achieved since 2001 – establishing democratic institutions and ministries being one of them. Primary education has expanded, roads have been laid and transport infrastructure has drastically improved' (Ahmed 2019).

With insecurity and conflict continuing and uncertainty regarding the willingness of major donors, particularly USA, to continue to support development assistance, the gains that have been made seem fragile at best. Over $US 70 billion (at 2017 prices) has been disbursed in ODA to the country between 2000 and 2017 and Afghanistan stands as the largest single aid project of the first two decades of the 2000s. It may yet be seen as aid's greatest failure.

Secondly is the issue of fiduciary risk, that is the risk that aid money spent and travelling through government systems may not reach its target and achieve its objectives. Much is said about corruption in developing countries (Rimmer 2000; Kramer 2007; Masoud *et al.* 2015), with accusations that politicians and officials find ways to divert aid funds to their own pockets or are able to influence donors to allocate projects to their own regions or ethnic groups (Briggs 2014). However, most donor agencies do not confront this issue directly, but instead focus on the much broader concept of fiduciary risk. There is a broad issue of fungibility (Howard and Rothenberg 1993) where aid received for some purposes, such as education, can allow governments to spend elsewhere, such as the military, but fiduciary risk is more complex. There are many other ways that aid funds may not reach their intended destination apart from overt corruption. For example, a government that operates under very tight budget constraints might find that it struggles to meet its everyday expenses (such as paying civil servants) when revenues do not come in as expected. If this happens and a large amount of aid funds appears, the Ministry of Finance may well be inclined to divert the aid funds to pay salaries and hope that other revenues improve so that the aid program can be recompensed in time.

Another issue has to do with dependence. With ODA being committed to large-scale programmes of expenditure, such as education, it can be difficult for recipient governments to find ways to expand their share of expenditure and eventually take the place of donors. Donor funding can become a semi-permanent feature of key government functions and the loss of donor funding can have catastrophic impacts.

Donors also struggle with the question of state legitimacy (Buiter 2007). When large amounts of ODA are committed to supporting the functions of the state, there is a strong indication that donors approve of the state and its leadership and recognise its legitimacy. However, as often happens, states are open to opposition and groups within a country may question the state's right to govern. This happens when the rights of regional minority groups are not seen to be recognised sufficiently and separatist movements arise, there are human rights violations by the state, or when unconstitutional or questionable methods are used to gain power etc. Donors are in an invidious position, for if they continue to channel ODA through the state they are seen to be supporting the government over opposition groups, and if they withdraw aid, they lose the opportunity to fund substantial programmes of change and are seen to be anti-government. These political considerations are crucial and often donors will decide on aid allocations on the basis of whether or not they want to fund a particular regime rather than necessarily whether the state has the ability to manage that aid effectively.

Aid for government functions also has some potentially negative aspects for recipient states themselves. Loss of independence inevitability follows agreements that involve large sums being tracked through state coffers. Donors understandably want to see that aid is well spent and want the recipient government to be accountable, but this requires a high degree of conditionalities over how internal systems work, what reporting will take place and the funding priorities of the state. There is a clear trade-off for recipients: accepting large aid donations means compromising the ability to manage the state completely independently. This leads to what we have called elsewhere the 'inverse sovereignty effect' and is particularly salient in smaller countries (Murray and Overton 2011b).

There are also more mundane problems for recipients. Despite the rhetoric of the Paris Declaration, donors have often proved slow to move towards predictable multi-year funding commitments. In practice, many still renegotiated funding on a year-to-year basis and the vagaries of donor budgets often resulted in volatile and unpredictable funding streams recipients. The example of the small Caribbean island state of Granada (Box 6.2) provides an illustration of how this country, dependent on ODA for a significant share of its development budget, has had to face large swings in aid flows from year to year, posing major difficulties for its financial management and forward planning.

Box 6.2 Aid volatility: the case of Granada

Granada is an independent state in the Caribbean Ocean. It has a population of approximately 112,000 people, mostly living on the main island. Its economy is based on tourism and some agricultural exports, notably nutmeg. The territory was under British rule from 1763 until independence in 1974. It remains a member of the British Commonwealth.

Aid has been an important source of income for Grenada at various times, but this has varied greatly over the past 45 years (Figure 6.2 shows aid inflows adjusted for inflation over time). Despite its status as a former British possession, the United Kingdom has not been a significant ODA donor; rather Canada has been the largest bilateral donor, though its role has been uneven and it seems to have reduced greatly in recent years. For the past 15 years, the World Bank as a multilateral agency, has been the main source of ODA for the country.

However, what is most noticeable in the profile of Granada's ODA over time has been its volatility. To some extent this can be explained by natural and political emergencies. In October 1983, USA led an invasion of the country in response to the coming to power of a perceived pro-communist regime there, but in the face of

Figure 6.2 ***ODA disbursements to Granada 1975–2017 ($US mill current $2017)***

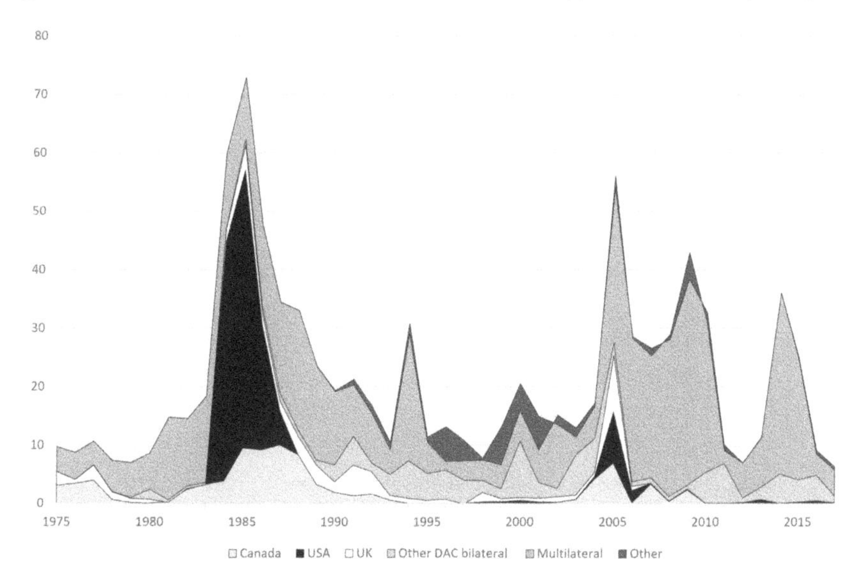

Source: www.stats.oecd.org

international condemnation. We can see that the invasion led to a large increase in ODA in 1984–87 notably from USA. But then the USA all but disappeared as a donor. Hurricane Ivan struck in 2004 and this led to another spike in aid (UK and USA contributed alongside Canada).

However, these two events cannot explain the other spikes and troughs in ODA flows to Granada. In the 1990s, there was an IMF-led structural adjustment programme put in place which led to cuts in aid and a reduction in the public sector. At other times, no doubt there have been donor perceptions of political instability or mismanagement and this has led to a reluctance to commit to ODA programmes. With the World Bank now as a major donor, volatility has not lessened, with low levels 2012 being preceded by relatively large inflows and followed by a peak in 2014.

Unfortunately, this picture of uneven aid flows is not unusual for small island states. Although natural disasters may introduce an element of uncertainty (Grenada is actually less prone to hurricanes than many other Caribbean states), the swings in aid receipts present particular difficulties for smaller economies. Planning and economic management is difficult, local financial and bureaucratic systems can become stretched some years, then under-utilised the next. Swings in ODA flow through to economic uncertainty and indecision. In these circumstances, aid volatility is not just a reflection of donor assessments of local governance performance, it can be a contributor to poor governance and bring negative economic consequences.

Recipients have also had to face what might be termed the 'burden of consultation' and lack of co-operation amongst donors. Following on from their 'good governance' concerns, donors have frequently, and rightly, attempted to ensure that there is wide public consultation regarding aid programmes and projects. This involves frequent public meetings, continual engagement with relevant government agencies, regular reporting and thorough monitoring and evaluation. Furthermore, each donor often has their own consultation requirements and procedures to follow in order to report back to their own agencies. Add in the large number of non-governmental organisation (NGO) agencies, the range of issues covered (from climate change to community policing) and the often-stretched recipient government and NGO institutions, and the result is often very high compliance costs for recipients to meet and report to a range of donors. For small island states such as Tuvalu, with a very small bureaucracy, the need to comply with global standards for consultation and reporting, the burden of consultation is heavy indeed (Wrighton and Overton 2012). Finally, for recipients, the requirement to adopt donor systems and processes may sometimes seem not only burdensome but inappropriate. Donors adopt their own ways of operating usually developed from their own social and cultural context. But these may not be suited to societies where, for example, there are strict social and political hierarchies and open public meetings may be dominated by certain elites. And for small island states or states with limited institutional capacity, large and complex international accounting and reporting systems may simply be too large and unwieldy (Overton *et al.* 2019).

These concerns have led donors and recipients to develop particular practices to mitigate some of these issues. The DAC-sponsored high-level forums on aid effectiveness (from Rome in 2003 to Busan in 2011 – see Chapter 3) aimed to develop guidelines and share good practice. Thus, the principles of aid effectiveness involved aspects such as donor harmonisation, use of recipient systems and so forth and there was much emphasis on improving financial management systems.

More recently, new donor approaches have been developed to tackle the 'development-security nexus' in situations where conflict and weak local governance have required more direct involvement, sometimes alongside military intervention (see Box 6.3). Here we see peacekeeping leading to the securitisation of aid, so that aid appears alongside military operations to restore order, then help build more secure institutions and government capacity.

Box 6.3 The Danish Peace and Stabilisation Fund

The Government of Denmark is an established ODA donor and historically its levels of aid have been relatively high for the Organisation for Economic Cooperation and Development (OECD), consistently achieving the 0.7 per cent target of ODA to Gross National Income (GNI). Much of its aid is directed to Sub-Saharan Africa. In 2010, and as part of the Danish Defence Agreement of 2010–14, it established the Peace and Stabilisation Fund (PSF), a comprehensive approach to dealing with conflict and security. Similar initiatives have been put in place in the UK (with the Conflict, Stability and Security Fund (CSSF)) and is being considered by USA (Price 2019).

The PSF constituted a new model for Danish aid. 'It is a joint, cross-government, funding pool to support multilateral and bilateral initiatives at the intersection of development and security in support of stabilisation and conflict prevention activities in fragile states' (Ministry of Foreign Affairs of Denmark 2014: 15). In practice, this has meant that resources from the ministries of justice, defence and foreign affairs have been brought together to constitute a fund to support efforts to promote peace and security in locations such as Afghanistan and the Horn of Africa. It is an example of a 'whole-of-government' approach to assistance overseas. Although much of the expenditure involves military operations, approximately 50 per cent of the PSF was counted as ODA between 2010–14, when total expenditure on the Fund was DKK 930 million (about $US 137 million) (Government of Denmark 2013: 23). By 2019, the Fund had increased to about $US 70 million per year, 80 per cent of which counted as ODA (Jensen 2019).

The Fund tackles the very difficult issue of development in conflict environments. It recognises the need for military intervention to ameliorate conflict, but also the need for development activities to help build economic activities and institutional capacity in order to maintain peace. Activities such as military operations and assistance for the strengthening of local military forces cannot be counted as ODA, but many other activities can.

The Fund is largely viewed as successful, although evaluations have called for better co-ordination (Ministry of Foreign Affairs of Denmark 2014). It is acknowledged that this comprehensive approach presents some difficulties in terms of focus and co-ordination: 'Military efforts aimed at eliminating immediate threats to stability and civilian interventions aimed at capacity-building often point in different directions when it comes to working with local actors' (Price 2019: 6).

These issues regarding mixed approaches may be mirrored by a degree of mixed motivations for this form of aid. In 2019 the PSF began a new programme in the Gulf of Guinea. As with some of its focus in Somalia, there was a concern to tackle the problem of piracy. The PSF has involved working with authorities in Nigeria and Ghana to address maritime security and crime. This has involved both anti-piracy operations, but also efforts to improve local legal and administrative systems. Whilst undoubtedly aimed at improving the security of maritime trade in the region, important for West African economic development, it has also been noted that some 40 per cent of seaborne trade in the Gulf of Guinea region is operated by Danish shipping companies (Jensen 2019). Clearly, peace and security operations are designed to benefit not only West African economies but also Denmark's own commercial interests in the region. Again, we can see how aid is becoming an increasingly complex phenomenon, moving beyond traditional notions of ODA and beyond established aid agencies to involve networks of agencies, combinations of funding sources and parallel interests of donors and recipients.

It is hard to quantify and analyse how effective aid has been in improving governance and government services in recipient states because improvements, if they occur, are highly qualitative: more transparency, greater confidence in government, more able officials, etc. However, it is clear that public sector change in many countries, following a decade or so of the neostructural aid regime, has been profound and largely beneficial. Monitoring of the Paris Declaration goals for a number of years showed appreciable increases in the use of local systems, moves towards untied aid policies. However, the targets set were frequently not fully achieved and, after 2008, the shift to a retroliberal aid regime seemed to take donor attention away from these state-centred performance measures. In other ways, though, the good governance and neostructural approaches did bring about important changes. Public sector reforms occurred across the developing world and many new and improved ways of operating were introduced whilst government programmes in health and education were bolstered considerably.

The relationship between aid effectiveness and governance is complicated by the fact that poor governance is strongly correlated with poverty, so that aid is more likely to go to states where good governance structures are in place and are thus more likely to be effective, whereas it is difficult to get good results from aid when poor governance is in place (Denizer *et al.* 2013; Levy 2014). Aid, however, along with other diplomatic measures, has helped bring about 'regime change' in places such as Kenya and Fiji where regimes with questionable democratic credentials were 'encouraged' to move to democratic elections and restore human rights. There seems to be evidence that aid has had a positive effect on promoting democracy, particularly in the post-Cold War era (Dunning 2004; Kersting and Kilby 2014). On the other hand, there are also cases where the withdrawal or non-existence of aid has not had a similar effect (North Korea and Zimbabwe, for example). As such it is hard to truly analyse the impacts on governance.

Thus, we might come to a tentative conclusion that aid can work to improve governance and government services, with long-term benefits for development under certain conditions (Dijkstra 2018). Carefully designed ODA programmes that conform to aid effectiveness principles can work with efficient state agencies to provide improved services for citizens and these, in turn, can provide the means for long-term improvements in well-being and widen the range of development options for individuals and states alike. Yet, when implemented less carefully – and perhaps when guided by motives other than aid effectiveness – support for recipient governments can bring about negative consequences in the form of supporting inefficient bureaucracies or undemocratic governments. And from the recipient side, changes in development management can lead to a situation where local control is compromised and governance is more a matter of pleasing and aligning with donor interests and ways of operating than forging strong and independent local development systems and strategies (Overton *et al.* 2019).

Aid and poverty

The rhetoric of poverty alleviation lies behind much of the public justification for aid programmes. Relieving human suffering provides the apparent *raison d'être* for many of the aid-funded projects and programmes world-wide and campaigns to increase aid, such as 'Live Aid' or 'Make Poverty History' appeal to this goal. Most aid and development strategies have poverty alleviation as an explicit objective, whether straightforward humanitarian relief, focused provision of basic services for the poor, or even economic growth models which promise a trickle-down effect for all (Collier and Dollar 2002). However, despite a basic view that the needs of the poor can be met by gifts from the wealthy, there is still much debate whether aid is very effective at alleviating, much less eliminating, poverty.

On the positive side, aid to alleviate poverty is largely based on a 'deficit framing' of poverty: it focuses on what poor people do not have and what they need to become better off. A lack of resources (whether food, clean water, good sanitation, basic education, health services, technical skills, equipment or financial capital) is seen as the basic obstacle to improvement in livelihoods and well-being for the poor and these resources can be provided from aid donors. This approach is most explicit following natural disasters or conflict when people's very survival is threatened by lack of food, water or shelter. Aid agencies, domestic and international and including governments, have proved able in most cases to identify these needs and respond accordingly (even if not as rapidly or completely as some would want). Aid has also been seen as successful in tackling major health problems, such as the incidence of malaria or smallpox (Levine 2007).

Perhaps the most prominent poverty-focused 'project' to tackle poverty was the Millennium Development Goals (MDGs) of 2000–2015. These, and the Sustainable Development Goals (SDGs) which followed for 2016–30,[1] identified some key dimensions of poverty though they tended to highlight the expressions of poverty rather than their underlying historical and contemporary structural causes. Thus, poverty alleviation was seen to be about raising incomes, improving maternal and child health, increasing literacy, ensuring women and girls achieve great equity of access to education and health etc., tackling major diseases, and improving sanitation, water supplies and housing. The MDGs and SDGs have therefore provided a substantial and explicit poverty agenda for aid. Aid should be directed to these key aspects of human well-being so that millions of people can move out of poverty.

Not everyone agrees with this approach though and critics argue that aid is not effective at tackling poverty at all. Neoliberal thinkers continue to argue that a free market is the best way of achieving economic growth and employment generation. They use metaphors such as 'trickle down' or 'a rising tide floats all boats' to suggest that market-led growth is more effective at dealing with poverty aid- and government-led interventions. Intervention creates distortions and disequilibria that lead to less than optimal allocations of resources (see the discussion on the economic impacts of aid above). Others, such as Easterly (2006), suggest that aid, despite the rhetoric of poverty alleviation, does not reach the poorest, but is instead captured by elites or squandered by inept donors or recipient governments. In some ways, these latter critiques hark back, albeit from a different ideological standpoint, to Robert Chambers' early (1983) criticisms of development practices ignoring the poor. If aid does not reach the poor, then it can contribute to rising, not falling, levels of inequality.

These debates are not easily resolved either way by the available evidence. Firstly, though, we can look at the record of the MDGs, recognising the very large aid commitments that were behind efforts to meet the goals. Although there were disappointments with the MDGs with most developing countries being unable to meet all the goals, there were some significant improvements. Access to health services improved markedly in many countries, as did participation in primary education. There is strong evidence that aid has helped lower rates of infant mortality, for example (Kotsadam *et al.* 2018). Gender inequalities were given attention as perhaps never before and, overall, the lives of many millions of poor people improved over the span of the MDGs. However, it is not clear how much credit aid can claim for these achievements. In particular the massive improvements on a global scale in incomes for many – the number of people living in extreme poverty globally fell from 1.9 billion in 1990 to 836 million in 2015 (United Nations 2015b: 4) – was largely due to the rapid growth of the Chinese economy, which had little or nothing to do with aid (Sachs 2012). Measuring the impact of aid is a counterfactual problem – we simply do not know what would have happened without it. It is also a multi-faceted and complex equation where it is difficult to separate cause and correlation.

So we are left with similar conclusions to those of Robert Cassen's studies of 30 years ago (Cassen 1994) – that aid can improve the lives of the poor, but it can also miss them out and, in the worst cases, make their position worse off. Many poor people do benefit from aid through improved access to schools and health services and some experience higher consumption levels and, through infrastructural improvements, better access to markets and information. Rural electrification can make major differences to the lives of poor households, even if it means that children can do homework by electric light at night, and better water reticulation similarly helps by ending the arduous task of fetching water from afar, usually undertaken by women and children. And it is beyond doubt that the lives of thousands have been saved by effective humanitarian interventions and assistance. These are real and tangible improvements in human livelihoods. Yet there is a nagging doubt that, despite the rhetoric, much aid does not reach the poor (for example, Briggs 2018) and has not been focused on poverty alleviation. Riddell (2007) suggested in 2007 that 40 per cent of the world's aid was not directed to the MDGs and this is almost certainly the case, if not worse, in the retroliberal period.

To draw lessons from these debates, we might suggest that aid can and does help tackle poverty. Humanitarian relief will always be needed in desperate times of need but when aid is directed to addressing the underlying causes of poverty, not just symptom relief, then long-term and sustainable improvements can be made. Better education and health services remain critical priorities for many countries and a focus on gender equity seems to bring substantial gains to welfare (Nussbaum 2000; Schultz 2002; Unterhalter 2012). Yet these are very long-term processes of change that require sustained aid support over perhaps generations, and they defy quick results. Attitudinal changes and societal transformations lie at the heart of deep-seated strategies to eliminate poverty, rather than a simple transfer of resources and filling of short-term need. Furthermore, as well as moving from short-term and quantitative approaches to poverty alleviation, aid programmes should also move from a narrow deficit-based framing of the issue to consider a more asset-based approach, building on, working with and respecting local assets, aspirations and knowledge.

Box 6.4 Oxfam and aid successes

In 2010, Oxfam produced a report *21st Century Aid* that amounted to a defence of aid and a challenge to increase commitments and improve practices. The report argued:

> *Aid plays a role in saving millions of lives. Recently, a barrage of criticism has been unleashed on aid, with critics using individual examples of failed aid to argue that all aid is bad and should be reduced or phased out altogether. This is both incorrect and irresponsible.*
>
> (Oxfam 2010: 1)

A catalogue of aid successes was presented. Examples included:

- 'There are 33 million more children in the classroom, partly as a result of increased resources to developing country governments over the past decade from aid and debt relief'.
- 'There has been a ten-fold increase in the coverage of antiretroviral treatment (ART) for HIV and AIDS over a five-year time span'.
- 'In Rwanda... [budget support] allowed the government to eliminate user fees for primary and lower secondary school education, increase spending on treatment for people living with HIV and AIDS, and provide agricultural loan guarantees to farmers.'

- For Tanzania's Local Government Reform Programme: 'The money provided by donors through the programme has funded over 4600 locally implemented projects including the construction of classrooms, roads, and clinics between 2004 and 2007, at the same time as building the capacity of the local government authorities involved, improving councillor and citizen involvement in planning and budget processes, and reducing transaction costs for service delivery'.
- 'In Zambia, recruiting and training large numbers of community-based health workers to distribute bed nets and safely diagnose and treat patients free of charge, in addition to indoor spraying, has reduced malaria deaths by a staggering 66 per cent over the last six years. The same approach has halved malaria deaths in Ethiopia in just three years'.

Overall the report was critical of trends in aid at the time towards diluting the poverty focus of aid: 'Aid that does not work to alleviate poverty and inequality – aid that is driven by geopolitical interests, which is too often squandered on expensive consultants or which spawns parallel government structures accountable to donors and not citizens – is unlikely to succeed' (Oxfam 2010: 2).

Aid and non-state sectors

Aid, particularly ODA, is frequently seen as working through a relationship between donor agencies (bilateral or multilateral) and recipient governments. However, non-state actors are often involved in various ways on both the donor and recipient sides (Wallace *et al.* 2007). The two key sectors here are civil society and the private sector.

The involvement of non-state agencies is important for a number of reasons. Firstly, they operate as substitutes for the state when recipient government agencies are regarded as inefficient, risky or not legitimate. Donor agencies can then by-pass the recipient government and not have to be seen to support it overtly. Thus, we see donor agencies turn to civil society agencies as partners in delivery of aid projects when governments are accused of being undemocratic or abusing human rights or excessively corrupt. Secondly, civil society and the private sector may be more flexible and cost-effective parties through which to distribute aid. Development NGOs often run with small overheads or rely on voluntary labour and private companies have other operations on which to fall back on. They can both pick up contracts for aid projects and thus have fixed and manageable involvement, preferable in some ways to maintaining a large permanent establishment to disburse aid within a government donor agency. With many NGOs and companies vying for contracts, donor agencies may find they can get good results for less money by using these agencies. Development NGOs and private firms thus become sub-contractors for aid delivery, tied to the donor agencies through legal contracts (Choudry and Kapoor 2013). Thirdly, non-state actors (especially NGOs) may be regarded as having closer ties to communities and poorer groups in society than governments. Many NGOs owe their origins to community organisations and draw their membership from communities. They are thus well placed to understand the needs of poorer communities and develop appropriate strategies for implementing aid projects. Civil society organisations can therefore be more responsive and better informed about poverty on the ground in remote rural regions than government agencies in a far -off capital city. In addition, donor development NGOs become part of this picture, for their relationships with recipient NGOs (some may be branches of the same international NGO) provides an effective set of networks through which a large number of hopefully well-informed and appropriate community projects can be supported efficiently by donor governments. These networks also often involve long-term durable relationships between individuals and agencies (also involving the development of deep level knowledge and understanding of conditions within recipient communities) that can be drawn upon to provide information or broker new projects when needed. These forms of social capital are extremely valuable in the aid world but are not always well recognised.

On the donor side too, there are good reasons to involve civil society. Development NGOs provide a key role in promoting development issues in the donor community. They help raise public awareness about poverty and development issues and, whilst this is primarily in order to raise funds to support their own operations, it can help donor governments who need to justify their aid budgets. Of course, this can also be a two-edged sword, for development NGOs will also seek to lobby governments to increase budgets more than they wish or change their aid policies. In many ways though, partnerships form between donor governments and NGOs and this was the case during the neostructural era of the early 2000s when both sides were committed to the MDGs: donor governments kept the NGOs on-side by drawing them into their aid programmes and offering contracts and funding. It had also been the case in the earlier neoliberal phase when donor governments sought to by-pass state agencies and use NGO contracting to fill the gap in, for example, limited welfare projects.

Private companies in donor countries have had an increasing role in aid in recent years. Part of the reason for this is shared with civil society: they may be flexible and cost effective in aid delivery through fixed projects and they can be drawn into supporting wider government aid strategies if they receive some of the funding. However, the reasons for involving the domestic private sector in aid is also so that donor governments can be seen to be channelling some of the benefits of large aid budgets back to the local economy. In the retroliberal era this has become much more common. There are aspects, such as the increased use of tertiary scholarships, which involve large portions of the aid budgets being spent in supporting donor tertiary institutions through fees and the domestic businesses through accommodation and living costs going into the donor economy. There is also the support of consulting firms which manage aid projects and provide expert advice to recipient countries but whose fees are received and spent largely in the donor country. Increasingly now, the involvement of private sector donor firms in aid is even more explicit with aid projects being designed to complement and help promote the offshore business operations of these firms. This is a far cry for non-state actors from using NGOs as cost-effective agencies to help deliver services.

Despite these advantages of using non-state agencies, we should also raise some questions and doubts about their use. These questions are rather different for civil society as opposed to the private sector. Civil society organisations are often regarded as efficient, well-informed, cost-effective and a good alternative to large, inefficient and corrupt state institutions. However, in practice, some development NGOs in recipient countries are not. Just because they are local does not mean they are any closer to understanding community conditions than, say, a local government official. They are often run by urban, relatively well-educated people. Many, yes, are based on real concerns for justice and poverty and are run by passionate, selfless and able staff but some are little more than middle class business operations, seeking to capture some of the revenue flows coming from donor agencies. Similar things might be said about development NGOs in donor countries. They are based on real concerns for human well-being and inequality and to address human suffering but, in order to function effectively, they have to run as businesses. The have to raise revenue to employ staff, pay rent and continue to support their offshore operations. In doing so they may appeal to basic sentiments of need and helplessness and use methods such as child sponsorship which may be effective at raising money from a sympathetic public, but which may well misrepresent the real conditions of people and wrongly convey simple pictures of complex development conditions and processes. Nor is there any guarantee that civil society organisations will be automatically any less inefficient, corrupt or ill-informed than their government counterparts. Indeed, some have to run with relatively high overheads and, due to small-scale and limited budgets, simply do not have the resources to put in place complex financial management and auditing systems (thus they find it hard to comply with strict donor conditions on such things). Finally, NGOs may have a difficult relationship with the state in recipient countries. Although involved in development work, many NGOs also are involved in advocacy and this can bring them into disagreement with government particularly when they suggest that government policies are part of the cause of developmental problems at the scale of the community and above. As a result, NGOs may be limited in their operations, confined to roles that are politically acceptable (so they can get funding and be allowed to operate), but perhaps staying away from important but contentious development debates.

For the private sector, the criticisms are rather different, but also considerable. Private businesses have to turn a profit to survive and this is both an advantage (they have to closely manage costs and revenue and run efficient systems) and a disadvantage (they will steer away from worthy development projects if they are not seen to be profitable). Businesses can operate well within fixed project cycles and with pre-defined and manageable activities and outputs but they are less likely to invest in long-term relationships unless there is a promise of financial return. Longer time horizons pose difficulties as does working in complex and unstable social and political environments (NGOs can be much more adept in these circumstances). Finally, and fundamentally, the private sector is in business to run business: being involved in poverty alleviation schemes and the like is a way of making a profit, not the primary goal. This said, there are many examples of businesses that run as social enterprises – they have social goals as their prime aim and, whilst still needing to earn a profit to continue in business, pure profit maximisation is not their fundamental objective (Kumar 2019). Much more than civil society, private enterprises are unlikely to become involved in political debates regarding the underlying or even short-term causes and impacts of poverty and underdevelopment. Economic activity rather than social and political transformation is their obvious focus.

To summarise, the involvement of both civil society and private sector companies has been common in different ways in the aid world for many years. They offer particular advantages and seem to be successful in contributing in ways that mainstream development agencies may find helpful at various times. They help make aid work. However, neither offers a clear alternative to state agencies and both have disadvantages and limitations. What is clear is that the NGOs and the private sector will continue to be important features of the aidscape, variously working separate from, but mostly alongside and with, both donor and recipient state agencies. The relative role is likely to be a political question and ebb and flow as aid regimes continue to evolve.

Aid and public accountability

Finally, in considering whether aid works, we now turn to consider whether aid works in terms of meeting public expectations regarding its objectives and openness. This aspect is not so much focused on the effectiveness of aid with regard to intended beneficiaries or the extent to which it aligns with principles of good practice (Knack *et al.* 2011). Rather, it has to do with the way aid agencies operate and report back to donor taxpayers and voters. In this way it is important to recognise that whilst donors frequently impose conditions on recipients to be transparent and efficient in the way they receive and manage aid receipts, the same conditions are not always imposed on doors themselves. Here we examine two examples of the way aid agencies have received recent scrutiny from independent organisations: the Aid Transparency Index and the Principled Aid Index.

The Aid Transparency Index is produced by the 'Publish What You Fund' campaign for aid transparency. This was launched in 2008 and receives funding from UK aid and the European Union. The campaign calls for aid donors to publish fully, openly, proactively and comprehensively all information on aid. It links to the International Aid Transparency Initiative (IATI), which has established standards, rules and formats regarding aid transparency. IATI maintains a register and tracks over 1,000 organisations.

The Aid Transparency Index is compiled annually and collects data on a variety of indices relating to budgets and finance, project attributes, 'joining-up' development data, performance, and organisational planning and commitments. It examines 45 major aid institutions including donor aid agencies (e.g. Department for International Development (DFID), United States Agency for International Development (USAID)) and multilateral organisations (United Nations Development Program (UNDP), World Bank). Each is ranked according to the results and categorised from 'very good' to 'very poor'. The 2018 results are depicted in Figure 6.3.

Figure 6.3 ***The 2018 Aid Transparency Index***

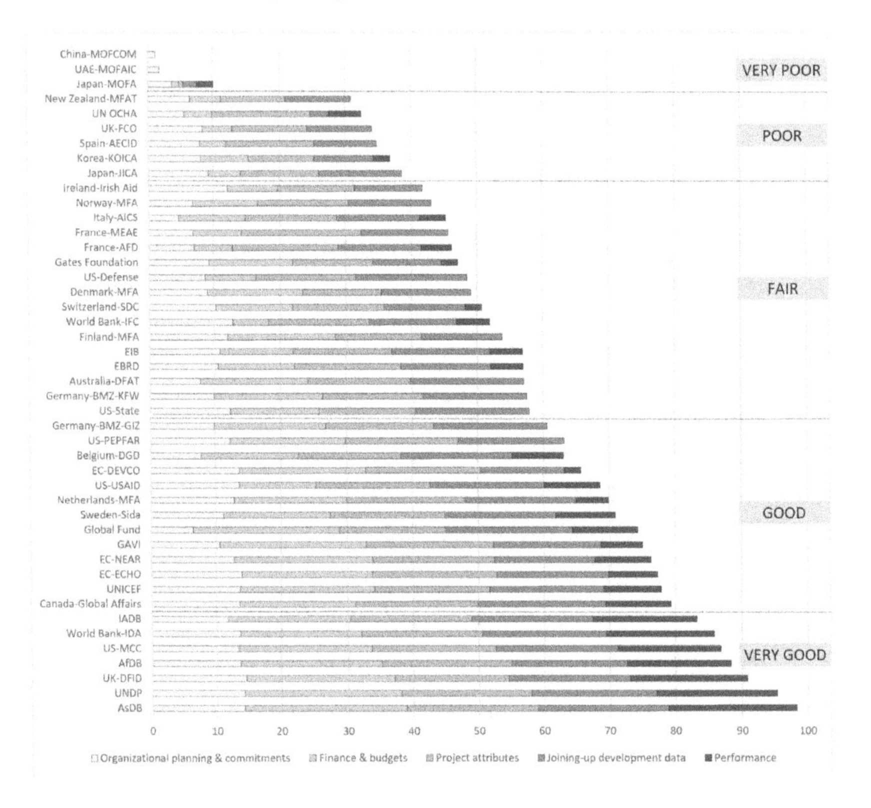

Source: redrawn from https://www.publishwhatyoufund.org/the-index/2018/

These results show wide variation in practice across the agencies. Some rank highly with close to full transparency. Notably this group includes the major financial agencies (Asian Development Bank, African Development Bank, World Bank International Development Association (IDA)). At the other end of the spectrum in the 'very poor' category are the foreign affairs agencies of Japan, UAE and China. Also of note is how different agencies from the same country score rather differently: UK's DFID ranks third from top but its Foreign and Commonwealth Office is sixth from bottom. Overall, the Index notes some improvements with more agencies publishing data on a regular quarterly or monthly basis using the IATI standard. However, there are still major concerns:

- 'More than a quarter of organisations do not provide descriptions of their projects at all or the descriptions provided cannot be understood by non-experts. Nor do all organisations regularly update datasets with accurate dates or provide the most up-to-date documents' (p.8).
- 'Some major international donors are not pulling their weight … This includes organisations at the bottom of the Index: Japan's Ministry of Foreign Affairs (Japan-MOFA), the Chinese Ministry of Commerce (China-MOFCOM), and the United Arab Emirates' Ministry of Foreign Affairs and International Cooperation (UAE-MOFAIC). These have not joined IATI but also make very little information publicly available elsewhere, despite being among the largest international donors' (p.9).
- At a time when ambitious global development goals, such as the SDGs, have been agreed and donors are increasingly under pressure to demonstrate that budgets are being spent effectively and have impact, the 2018 Index … reveals serious data shortfalls (p.25).

Overall, we might add that not only do most donor agencies need to improve their performance with regard to the transparency of their aid policies, expenditures and evaluations, they also many need to practice better themselves what they require of their recipient partners.

The Principled Aid Index is produced by Overseas Development Institute (ODI) in the UK (Gulrajani and Calleja 2019). It focuses on the 29 DAC donors and uses data on their aid spending to analyse the extent to which their aid can be considered 'principled'.

'Unprincipled' aid is regarded as 'self-regarding, short-termist and unilateralist. Donors concentrate on securing narrower commercial or geopolitical interests from their aid allocations while sidelining areas of real development need or undervaluing global cooperation' (Gulrajani and Calleja 2019: 2). The Index is formulated on the basis of three components (needs, global co-operation and public spiritedness), each with four quantitative indicators (Table 6.1).

Table 6.1 ***The Principled Aid Index: principles and indicators***

Principle	*Definition*	*Indicators*
Needs	Aid is allocated to countries to address critical development needs and vulnerabilities	**A. Targeting poverty**: Share of bilateral ODA/gross national income (GNI) targeted to least developed countries (LDCs)
		B. Supporting displaced populations: Share of ODA to developing countries that cumulatively host 70 per cent of cross-border forcibly displaced populations
		C. Assisting conflict-affected states: Share of humanitarian ODA to countries with active violent conflicts
		D. Targeting gender inequality: Share of bilateral ODA to countries with the highest levels of gender inequality
Global co-operation	Aid is allocated to channels and activities that facilitate and support global co-operation	**A. Enhancing global trade prospects**: Share of bilateral ODA to reduce trade-related constraints and build the capacity and infrastructure required to benefit from opening to trade
		B. Providing core support for multilateral institutions: Share of ODA as core multilateral funding
		C. Tackling the effects of climate change: Share of total ODA (bilateral and imputed multilateral) for climate mitigation and adaptation
		D. Constraining infectious diseases: Share of total ODA allocated to slow the spread of infectious diseases
Public spiritedness	Aid is allocated to maximise every opportunity to achieve development impact rather than a short-sighted domestic return	**A. Minimising tied aid**: Average share of formally and informally tied aid

		B. Reducing alignment between aid spending and United Nations (UN) voting: Correlation between UN voting agreement across donors and recipients, and donor ODA disbursements to recipients
		C. De-linking aid spending from arms exports: Correlation between donor arms exports to recipients, and ODA disbursements to recipients
		D. Localising aid: Share of bilateral ODA spent as country programmable aid (CPA), humanitarian and food aid

Source: Gulrajani and Calleja (2019: 3)

Each donor is scored (out of ten for each of the three principles) and ranked. The results for 2017 are shown in Figure 6.4. As with the Aid Transparency Index, we see considerable variation across the range of donors. Not only does the total score vary, but also donors may be unevenly strong or weak on different principles. Thus, Slovenia, for example scores poorly on global co-operation, but highly on public spiritedness. France is poor on responding to needs, but the second best at committing to global co-operation. Luxembourg, as the top-ranked donor, is strong in all three principles.

In tracking scores over different years, the authors of the index found that donors are becoming more principled in total but that there is 'a worrying deterioration in donor commitment to public spiritedness' and 'many donors are adopting a more short-sighted approach to aid, targeting it to help domestic constituencies and firms and supporting short-term foreign policy objectives, rather than taking a longer-term, principled approach' (Gulrajani and Calleja 2019: 1 and 7). This finding is in line with our earlier analysis of the retroliberal aid regime in the past decade, putting donor self-interest ahead of poverty-related needs. Also worthy of note is that fact that, when the index is set alongside a measure of donor generosity (the ODI/GNI ratio), there is a positive correlation evident: the more generous donors are more likely to be more principled.

Along with these two surveys across the range of donors, there are some instances of agencies scrutinising particular country programmes. One interesting example is the UK's Independent

Figure 6.4 ***The Principled Aid Index 2017: scores by country***

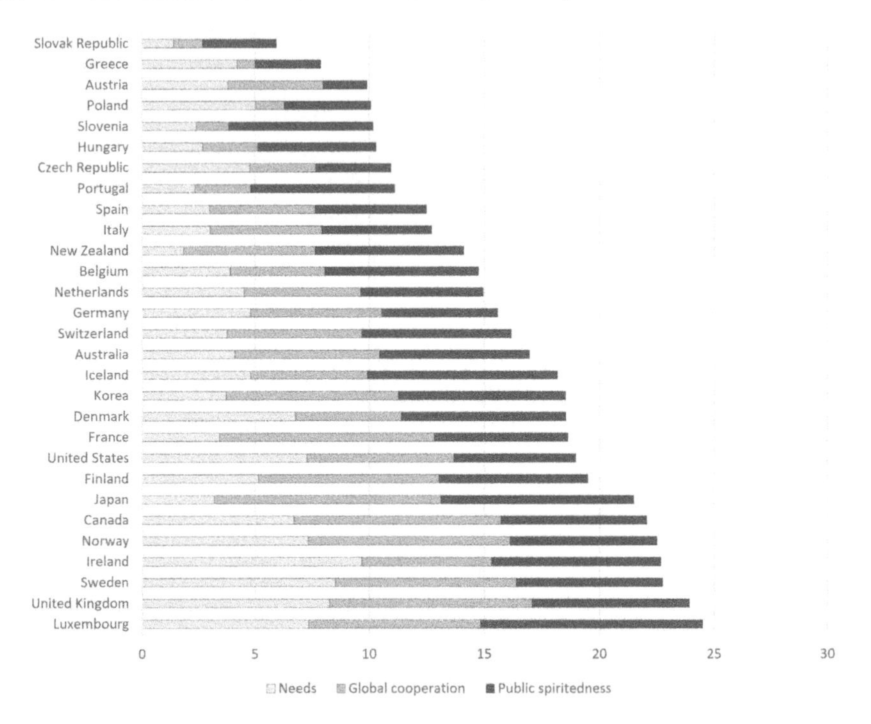

Source: redrawn from https://www.odi.org/opinion/10502-principled-aid-index

Commission on Aid Impact (ICAI). This agency was established in 2011 as a watchdog for British ODA. It operates independently of government and works 'to ensure UK aid is spent effectively for those who need it most, and delivers value for UK taxpayers' (https://icai.independent.gov.uk/about-us/). The agency produces reports on various aspects of the aid programme, whether overseen by DFID or other government agencies. For example, it has been critical of the business in development approach of DFID (ICAI 2015), and the investment arm of UK aid (CDC) (ICAI 2019a). An overall review of its 32 investigations between 2015 and 2019 (ICAI 2019b) recorded a variety of 'scores' (on a range from green to red, from effective to not effective). Of the 24 activities which were scored, only one (efforts to eliminate violence against women and girls) received a clear 'green' rating, whilst eight received a worrying 'amber/red' score. Overall, it raised concerns about the direction of the country's aid programme and cautioned that the UK's 'pivot back towards upper-middle-income countries does not lead to neglect of the key SDG objective of eliminating extreme poverty and inequality' (ICAI 2019b: 25).

These forms of scrutiny on donors are important and suggest that when we consider the question of whether aid works, we need to examine closely the performance of donors in formulating, targeting and reporting on aid, as well as analysing the impacts of aid on recipient economies, societies and systems of governance. Donors should be just as accountable for aid as recipients.

Conclusion

We started this chapter by asking whether aid works. Sadly, there is no clear answer, in part because it depends on how the question is posed. There is evidence and there are arguments to suggest that aid can indeed bring benefits in terms of economic activity, improved welfare, relief from hardship and even reduced levels of poverty. Aid has often worked, but on the other hand, it frequently has not. Aid can bring harmful effects: crowding out local resources and initiatives, producing dependence, being appropriated by the wealthy, and widening inequality.

We suggest that, although there are no clear and obvious answers, aid seems to be more effective when it is directed at working with and building local assets and capabilities, at providing the important elements of welfare services (health and education) and basic needs (clean water, sanitation, shelter) that allow people to survive, be healthy and then take steps to improve themselves, their families, and their communities. This is not to suggest that the state should effectively practice a totally hands-off self-help philosophy. The state must intervene to facilitate and support development through the correction of market failures and negative social externalities, which we argue are large and damaging in the case of developing economies. Aid plays a very important role in this regard. But it cannot do it alone, a strong and effective civil society is required together with an enlightened private sector. On the other hand, when aid becomes too involved in private sector operations, whether subsidising some elements and not others or taking the place of local enterprises, there seem to be questionable results and likely distortions. Aid can work well when in partnership with private companies and local government agencies and it can help support and build public capacity to undertake important development programmes, yet this is no automatic path to success for there are many pitfalls. For aid to be effective, we also contend that it has to be transparent and principled on the donor side.

However, to return to our question, 'does aid work?', perhaps we are asking the wrong question. We are judging aid by criteria that may seem to be intuitively valid – does aid promote economic growth, does it improve governance, does it alleviate poverty etc. – but which are not at the root of why aid is given in the first place. The bulk of global ODA is given by powerful Western countries who support aid budgets as part of their overall foreign policy strategies. Aid programmes are supported in a rhetorical sense by appeals to poverty alleviation, democracy, human rights, or humanitarian relief. Public support for aid rests largely on these explicit goals. However, aid budgets are only sustained because they help achieve larger foreign policy objectives: political stability, strategic alignment, economic integration and trade, and, overall, national self-interest. And although these things are almost impossible to quantify and analyse in a cost-benefit manner, it is apparent that aid does indeed work, for if it did not help achieve these wider goals, aid budgets would have been cut long ago! Nonetheless, we (as with the authors of the Principled Aid Index) argue that donor self-interest is in fact best served in the long-term by contributing to a 'safer, more sustainable and more prosperous world' (Gulrajani and Calleja 2019: 2; also Collier 2016), rather than the short-term pursuit of donor economic and political goals. This intangible but infinitely valuable prospect will yield long-lasting progress and prosperity for donors and recipients alike.

Summary

- There is a range of viewpoints concerning whether aid works or not. Opinions are often rhetorical in that they do not refer to empirical evidence. Furthermore, the answer to the question 'does aid work?' is dependent on how the question is constructed.
- A central objective of aid has been to promote economic growth. There are various models rooted in modernisation theory that argue the need for capital and technology. Critics from a neoliberal point of view have suggested this crowds-out, distorts prices and leads to the application of inappropriate technology.
- Neoliberals consider that intervention in the economy creates negative externalities whilst neostructuralists argue that negative externalities can only be solved through intervention.

- The most serious empirical studies are clear that long-term aid can work positively for the economy, under certain circumstances. There is no quick fix and broad programmes that incorporate infrastructure, improve human capabilities and technological progress work best.
- Beginning in the neoliberal aid regime there was a focus on good governance as a target for aid. Emphasis on 'rolling back' the state turned to facilitating the state in the neostructural period.
- The Paris Declaration of 2005 and the resultant focus on ownership and effectiveness greatly enhanced the use of aid towards capacity-building programmes with the long-term objective of facilitating governments to bring about progressive development.
- The evidence on the relationship between aid and good governance is mixed, Sometimes the concept of good governance has been used as a smokescreen to further donor geopolitical aims but sometimes it has precipitated positive structural changes in governance that have yielded positive outcomes for populations.
- Aid is often justified as a means of reducing poverty. However, this has not always been its true objective.
- Through the MDGs and the subsequent SDGs there has been an explicit concern with poverty reduction and latterly its elimination. However, reductions in poverty cannot be ascribed solely to the MDGs.
- Critics of the role of aid in reducing poverty often use anecdotal evidence to support their views. A more rounded empirical approach suggests that long-term policies that target the poor have a positive impact.
- Non-state actors have become increasingly involved in aid flows. Civil society and the private sector are seen as more flexible and 'less political'. However, there are drawbacks in the case of both.
- Donors need to be as accountable for aid allocations as recipients, as they clearly prosper from it and it helps justify it politically. There is a wide range of transparency between donors, although generally there is room for improvement.
- Overall, some aid has worked and some has not. Aid that aims for long-term, broad-based, inclusive development has been the most successful and will continue to yield results that are not instantly measurable but nevertheless enormously valuable.

Discussion questions

- What are the economic arguments for and against aid?
- 'Aid undermines the ability of recipient governments to manage their own affairs'. Discuss this statement in the light of arguments concerning aid and good governance.
- Is there a direct correlation between aid and poverty reduction? If not, why not?
- According to empirical studies what are the best approaches to aid to reduce poverty?
- What explains the rise of non-state actors in the aid sector and how effective have these groups been?
- When we assess if aid works, what measures should we use as evidence?

Websites

- The Aid Transparency Index: https://iatistandard.org/en/
- The Principled Aid Index: https://www.odi.org/opinion/10502-principled-aid-index
- ICAI: https://icai.independent.gov.uk/

Videos

- https://www.odi.org/publications/11294-principled-aid-index-understanding-donor-motivations

Note

1 The SDGs continue a concern for poverty alleviation – in fact this is extended to setting the ambitious objective of 'poverty elimination' – but are not solely about poverty as the MDGs professed to be. The SDGs include broader concerns for environmental sustainability, justice and inequality and widen the concern to all countries.

Further reading

Addison, T., Morrissey, O. and Tarp, F. (2017) 'The macroeconomics of aid: Overview', *The Journal of Development Studies* 53(7), 987–997.

Barcelos, P. and De Angelis, G. (eds.) (2016) *International Development and Human Aid: Principles, Norms and Institutions for the Global Sphere*. Edinburgh University Press, Edinburgh.

Easterly, W. (2003) 'Can foreign aid buy growth?', *Journal of Economic Perspectives* 17(3), 23–48.

Gibson, C.C., Andersson, K., Ostrom, E. and Shivakumar, S. (2005) *The Samaritan's Dilemma: The Political Economy of Development Aid*. Oxford University Press, Oxford.

Jones, S. and Tarp, F. (2016) 'Does foreign aid harm political institutions?', *Journal of Development Economics* 118, 266–281.

Oxfam (2010) *21st Century Aid: Recognising Success and Tackling Failure*. Oxfam Briefing Paper 137.

Riddell, R.C. (2007) *Does Foreign Aid Really Work?* Oxford University Press, New York.

7 Conclusions

Futures for aid

Learning objectives

This chapter will help readers to:

- Revise and recap the range of arguments concerning the definition, measurement and effectiveness of aid from a range of perspectives
- To make informed predictions with regard to what the aid sector might look like in the coming years
- To understand three views on the future of aid – pessimistic, realistic and optimistic
- To consider the possibility of aid beyond ODA

Introduction

In this book we have examined aid and the ways it is defined and distributed, how this has changed over time and whether or not it has been seen as successful and effective. We started by suggesting that there is no agreement about how it is defined, how it is measured and whether it works or not. There exists a wide range of views on all of these matters, from a variety of perspectives. We are now able to reflect on these various debates and look forward and speculate about how aid might change in the future. We begin by recapping and extending the discussion concerning effectiveness and then move onto explore a number of future scenarios.

Aid debates revisited

Debates about the effectiveness of aid covered a wide spectrum of views. There were those who generally felt that aid has the potential to bring about beneficial change to the lives of many millions of people by improving their access to resources and social services. It can help provide jobs, new goods and services, improved schooling and health services and better infrastructure. It can also assist developing country governments in augmenting what they provide to their citizens and improving the way they operate. Aid is also seen positively not just by governments and big international agencies, such as the World Bank, but it is also supported through donations by many people worldwide who see human suffering and feel they can help improve the well-being of distant others.

On the other hand, the critics of aid come from two main strands. On the political right are those who believe that aid does harm because it distorts economies and disrupts the incentives for poor people to work and determine their own life paths. Dependency, inefficiency, corruption and poor economic performance are the consequences of large amounts of aid given to governments. Conversely, on the political left, are those who argue that aid is little more than a tool of new imperialism, a mechanism for rich and powerful countries to exert their influence and control over struggling states and communities. They argue that aid misses its stated targets of helping the poor and instead serves the interests both of the donor side (both governments and big business) and the recipients (the elites and businesses tied to aid flows).

Our view in this book does recognise that aid can be poorly designed and distributed. It can indeed distort economies and be co-opted by elites and special interests. And it often fails to meet its objectives of poverty alleviation and economic development. However, we also recognise that there are many ways in which in its various forms aid can, and does, work well. Humanitarian assistance, albeit with examples of failure, has helped bring relief to many who have suffered from natural and human-induced disasters. To end such forms of aid would be to magnify the suffering of many. Over the coming generations, with the certainty of increased climate change-related natural disasters and conflicts over water and other resources, such aid may be even more desperately needed. We are likely to see public appeals and government responses for relief aid for many years to come. We have also seen that some other forms of aid can work very well. Often, small projects – with aid inputs that build on local communities' needs, resources, knowledge and initiatives – can have a profoundly positive impact on the daily lives of people whether that be through ready access to potable water, the provision of a health clinic, a road or primary school, or training and support for a village-run microcredit scheme. However, we feel the greatest promise for positive aid outcomes rests in higher-order modalities and long-term commitments, particularly SWAps and GBS (see Chapter 4). These support state agencies to provide essential services to a wide population. It is these mechanisms, with substantial volumes of aid backed by trust in local agencies to plan, manage and allocate resources, that have the greatest potential to bring improvements on a large scale. This approach was endorsed strongly in the early 2000s, backed by the Millennium Development Goals (MDGs) (and later Sustainable Development Goals (SDGs)) and agreements such as the Paris Declaration of 2005, and seemed to represent a consensus at the time as to what constituted effective aid. In short, we are of the view that aid can be an effective tool in channelling resources from wealthy economies to those who need support to tackle widespread poverty and inequality if effectively conceptualised, distributed, managed, and applied.

More broadly, we might also argue that the aid debates regarding the effectiveness or not of aid in promoting growth or alleviating poverty rather miss the point. Aid may appeal to the rhetoric of 'development' or 'humanitarian relief' or 'poverty alleviation', but it only survives because donors believe it is in *their* best interests to give aid. Aid is not a one-way transfer of resources but rather a means to secure political and economic position and strategic leverage for donors. It involves two-way exchanges, often transparent in terms of ODA flows from donors to recipients, but frequently opaque and varied in terms of reverse flows. So, in addressing the question of whether aid works or whether it is effective, perhaps we should just note that it *must* be – in meeting the broad objectives of donors that is – as long as they continue to fund it.

What is important though, aside from this rather cynical view, is that we continue to monitor aid negotiations, agreements and policies to help prevent abuses and the mistakes of the past. In addition, as students of aid and development, we should keep parties to account – to ensure that the putative objectives of aid they espouse (and justify it to their electorates) are kept in mind and that the best practices for achieving these goals are adopted.

Moreover, it is possible to suggest that there are ways aid can work effectively with regard to public rhetoric and support for it:

those objectives based on social justice, human rights, democracy and eliminating poverty. Aid, we argue, can and should bring demonstrable gains for the poor in terms of survival, well-being, security, justice, sustainability and widened options for future livelihoods.

Aid futures

Looking to the future, and knowing something of where we have come from, it is possible to speculate about what aid practices might emerge in coming years. There are already some signposts for these that we can extrapolate forward. Three scenarios are suggested here: a *pessimistic view* that sees the demise of aid as a progressive force for tackling poverty; a *realistic view* that notes the rising overlap between conflict, military intervention and humanitarian and development assistance; and a more *optimistic view* that sees in the SDGs the basis for a new global aid project.

A pessimistic view

Firstly, in terms of the *pessimistic view*, some commentators (Mawdsley *et al.* 2014) have pointed to the possibility of a 'post-aid world' or argue that aid is becoming less relevant and we need to look 'beyond aid' (Janus *et al.* 2014). There are certainly some strong indications that we are heading in this direction. Firstly, the important example of China demonstrates that some countries (on both sides of the aid relationship) prefer to see relationships framed in terms of South–South co-operation rather than North–South aid. China has been very active in providing grants and loans to build infrastructure throughout the world. And it has been quite open about the fact that it sees this as a way of promoting its own interests in trade and in employing its own companies to construct the roads, buildings and telecommunications it funds. Two-way flows and expanding economic relationships are explicit goals of this approach. China has also resisted the efforts of the DAC and Western donors to 'join the club' and work as they do. The continued rise of the South–South co-operation model seems to not only challenge the old North–South framing of aid, it is perhaps influencing a move away from our very notions of 'aid' (Mawdsley 2019).

Western donors have changed their approach. The retroliberal era, as we describe and analyse it, has seen a move away from the neostructural approaches of poverty alleviation and high-level modalities to more overt self-interest on the part of donors, greater support for their own business interests (as seen in emerging public-private partnerships and new models of funding for development) and the adoption of the mantras of 'shared prosperity' and 'sustainable economic growth'. Indeed, this represents a sort of convergence with Chinese practices and a divergence from the old models of aid based on supposed Western altruism and the grand narrative of poverty alleviation on a global scale (Mawdsley 2017; Overton and Murray 2018; Gulrajani and Faure 2019). These seem to be steering us increasingly towards a situation in which aid as we knew it is fading away, aid institutions and mission statements are being transformed and aid, if it survives, is seen as just another means for states to support market-led economic growth.

Furthermore, even this 'post-aid' scenario seems to be open to further change and retreat. With the election of more inward-looking, conservative populist and overtly nationalist governments in the USA and UK, together with knock-on effects across the world there is an emerging autarkic nationalism. UK and USA are two of the largest donors and this is likely to have an influence in real financial flows as well as policy direction. Conservative critics of aid may well be more in the political ascendancy and further reductions and restructuring of aid programmes are likely. Aid in this scenario is currently under threat.

A realistic view

A second scenario, referred to here as the *realistic view* involves the continuing involvement of aid agencies and the increasing allocation of aid funds to conflict situations. Military interventions and conflicts in Afghanistan, Iraq, Syria and other theatres has, as we saw in Chapter 3, brought large amounts of aid in their wake. Often in the past, conflicts zones and sites of violence were places where all but the hardiest and most specialist of aid agencies (such as the Red Cross and UNHCR (United Nations High Commission for Refugees)) would go. Military force was used to secure peace and then aid agencies would move in to tackle development issues, build prosperity and, hopefully, help prevent future conflict. In the past two decades, however, not only have there been signs of more assertive intervention by the UN and other forces, aid agencies have been part of continuing efforts to bring about change in regions of conflict. Military solutions alone have not been effective in bringing about peace where conflict is deep-seated and complex. Violence continues to occur, historical tensions bubble away and there is no clear break between conflict and peace. Aid work then is pulled into these situations, sometimes working uncomfortably alongside military forces (with whom they are seen to be associated and from whom they receive protection and logistical support). Military institutions themselves have learned that they too need to be involved in development work, working with communities, helping to provide needed facilities and building stable institutions. As we saw in Chapter 2, Official Development Assistance (ODA) has been quietly redefined to now include some forms of military expenditure. This scenario provides some difficult challenges for aid agencies. It takes them into situations where violence and security are critical daily considerations, it takes away their preferred status as politically neutral agents when they appear alongside military forces, and redefines their work so that development for peacebuilding is seen as a key objective rather than, say, development for poverty alleviation.

A related aspect to this second scenario is the increasing involvement of aid in refugee work. Aid has always had this connection and many agencies have worked with both international refugees and internally displaced persons in many different countries. Yet these refugee crises were usually regarded as distant and to be dealt with far beyond the borders of donor countries. In recent years, however, the rise of flows of political and economic refugees from Syria and North Africa into Europe, from South and Central America towards other wealthier countries in the region, such as Chile and Argentina, or various people trying to land by boat in Australia, for example, has led to a rethink (and redefinition) of the relationship between aid and refugees. Refugee crises now occur within or on the borders of donor states, especially in Europe, and the costs of protecting and accommodating these flows has also become incorporated into aid definitions and ODA accounting. We are seeing more ODA being spent within many donor countries as a result. As with conflict situations and work within military-led missions, this has implications for how we see and practice international development assistance. Interestingly, it brings aid into the very countries where aid budgets are generated.

An optimistic view

Our third scenario is a rather more *optimistic view* and sees a future for renewed and revitalised aid efforts directed to and meeting the real needs of the poor and marginalised. The replacement of the MDGs with the SDGs after 2015 seemed to signal a continuation of the global agreement to address issues of poverty, yet there were some important changes that came with the SDGs. Firstly, the goals were developed after a much deeper and longer process of consultation and discussion and, more than the MDGs, they had broad support with civil society and countries of the Global South much more involved in their formulation. The goals were larger in number (17 not 8) and complexity and moved beyond just poverty-focused objectives (income poverty, literacy, maternal health, gender equity etc.) to include a wide range of environmental and social justice issues, such as climate change, the health of oceans and peace and justice. And they appeared to be more ambitious: 'poverty alleviation' became 'poverty elimination' and 'inequality' appeared as a separate goal. Finally, the goals were designed to apply to all countries – gone was the old MDG model of the North–South divide where the goals were to be pursued in the South, funded to a large degree by the North. So the SDGs are ambitious, wide ranging and universal.

The SDGs are having and will continue to have significant implications for aid. Firstly, by being larger in number and more wide-ranging they may involve a more jigsaw-like approach to aid strategies. Countries will need to assess which goals are most appropriate for them and design development strategies accordingly. The old template approach of PRSPs to poverty alleviation, for example, may need to be replaced by more tailored national plans applied to a diversity of situations. Secondly, the applicability of the SDGs to all countries may further weaken the North–South imaginary that characterised the aid world of the past. Poverty and high inequality exist within wealthy countries, the majority of the world's poor now live in middle-income countries not just the poorest, and the threats to sustainable development have to be seen as existing within the Global North (through the production of greenhouse gases or unsustainable use of non-renewable resources) as well as the Global South. Future aid may well involve more inward-looking action and (for donors) uncomfortable self-examination. Thus, the SDGs could well revive the global commitment for action and the allocation of significant resources towards improving the secure and sustainable lives and livelihoods of the world's population.

However, given the fading of the North–South divide this may not involve a significant re-allocation of resources to the Global South. Change under this scenario will not simply be a re-creation of the neostructural era's poverty project. Rather it will be an explicitly global project aimed at ending poverty. It will need imagination, critical self-reflection and a substantial and sustained commitment of resources targeted in much more intricate and, hopefully, effective ways applied appropriately at various scales.

Aid beyond ODA?

Much of this book has focused on the type of aid most used and best understood on the global stage: ODA. This form of aid has been tightly defined, measured, monitored and analysed over the years. We have a fairly good picture of global ODA and it continues to dominate by far the flows from donors to recipients in the name of development assistance. Yet we also noted that it has itself been under review; its parameters and definitions have changed and the concept of ODA does not include many forms of assistance, quantitative and qualitative, that exist and have existed as part of the web of aid relationships. If, as above, we speculate that aid (and ODA in particular) is likely to change a great deal in the future, what can we say about forms of aid outside the ODA framework as well as how this might alter aid relationships and networks overall?

For a start, it is important to recognise that many people's understanding of, and involvement in, aid is not part of the ODA framework in the first place. When disasters strike, non-government organisations (NGOs) move into action, raise public funds and provide relief and assistance. Many development NGOs in Western countries exist explicitly in order to raise funds and distribute these overseas. They continue to spread awareness of issues of poverty and hunger and many are able to survive without government support. Importantly too, much 'disaster relief' comes through familial and informal networks so that diasporic populations will raise money to send to families in need. These remittance flows, in cash and kind, may well be small by global ODA standards but they are significant for those involved and they continue to operate, and will operate in future, largely separate from government-led efforts. A 'post-ODA world' might emerge as governments retreat from established aid activities but people's everyday understandings and involvement in assistance for 'others in need' continues, and in some cases grows to replace more formal networks. A 'post-ODA' world is different to a 'post-aid' world.

A critical form of assistance – labour migration – is also likely to continue and may even expand. We noted that migration and subsequent savings and remittances often brings substantial benefits to migrant-sending countries (despite some reservations and potential costs). Many developing country leaders are pressing wealthy countries to open their borders more to allow more labour migration to occur. At the time of writing, this has become a highly charged political point in a number of territories and is leading to considerable political shifts, especially to the right in the USA and Europe. This is not a form of ODA of course, but more liberal migration regulations can be considered a form of development assistance and migration schemes are frequently cast in development assistance terms. A receiving country's migration policies are set by an assessment of its own labour needs, social policies and political climate, not by the development benefits it brings to sending countries and communities. Thus, migration as a form of development assistance will continue and may even become relatively more important but it will be determined by the vagaries of both the global economy, affecting labour demand and supply, and internal domestic politics. On the other hand, resistance to immigration is also affecting aid policies in that ODA is being used more explicitly in attempts to shore up the economic and security contexts of migrant-sending regions. We argue that migration, and its relationship to development, needs to become a much more prominent element in thinking about aid.

The relationship value of aid

Aid – and ODA in particular – stresses material things, whether financial resources, equipment, consultants as advisors, construction and so on. But, as we have recognised, the motivations for aid are often associated with more abstract matters, tacit understandings, reciprocity, and relationships. And aid also has symbolic value in terms of the act of giving and receiving between the parties involved (Mawdsley 2012c). We are of the view that these 'relationship values' within aid exchanges are often understated and undervalued. Operating at different levels, they will continue to be important well into the future. For example, some aid agencies specialise in the sending of volunteers to work in developing countries. The volunteers get paid little and their costs may be very small compared to a private consultant sent to give 'expert' advice to recipients. The volunteers hopefully are of some use to their hosts, often bringing professional experience or technical skills when they work alongside counterparts within local organisations and communities. However, the essence of effective volunteer work is the relationships that are developed: volunteers on longer placements (of, say, two years) frequently develop friendships, trust, knowledge of the local environment and empathy. As a result, their work is more effective and they also gain from the placement in terms of personal satisfaction, life experiences and personal relationships. The personal also flows through to the institutional so that when volunteers from one agency continue to come to a host, long-term goodwill and understanding are built and mutual learning takes place.

At a larger level, person-to-person relationships are also crucial within and alongside aid. We have seen how aid is used to restructure recipient institutions so they align with donor systems and each can understand the other better; scholarships are also used to build relationships between members of future elite groups and the donor/host country; and aid helps cement diplomatic ties and relationships between government officials and community leaders. In this sense, there are strong reasons for maintaining aid as an important diplomatic tool. These reasons may also extend in interesting ways too. In August 2017, Hurricane Harvey hit Texas and caused large-scale floods with much damage and some loss of life. USA at that time was not on good terms with Venezuela with President Trump criticising President Maduro and even threatening military action. Venezuela, however, and as it had done in 2005 in the aftermath of Hurricane Katrina, offered an aid package to America of $US5 million (Erickson 2017). This may be regarded as a political stunt by some, but it is an important symbol that aid can run two ways, breaking down existing conceptions of wealth and power. In this regard it should be seen as a means to build positive relationships.

Towards more effective aid

Given these considerations regarding the importance of relationships in aid, and in looking back to our analyses of whether and how aid works (Chapter 6), it is possible to suggest some lessons we may have learned with regard to the way aid might be used to achieve more effective and positive development outcomes. These lessons are directed towards aid practices which prioritise the objective of improving security, well-being, justice and sustainability for the poor (Box 7.1).

Box 7.1 Lessons for effective aid

1. *Be attentive*: Aid priorities should be articulated by those who will supposedly benefit from assistance. Donors should not predetermine what is needed or lacking but instead listen and let local agencies, government and civil society define not only their aspirations but also the means for their achievement and the assets they have to work with for a start. Notwithstanding this, donors (as well as recipients) should exercise their right to say 'no' when they disagree (see (6) below).
2. *Be trusting*: Aid should be a way for local agencies to own, maintain and continue their own development strategies. This requires donors, once they trust those agencies, to give resources freely and without conditions so that activities will be pursued in a sustainable way. Donors should respect local ownership and align behind local leadership and within their systems.
3. *Be patient*: Fundamental development takes time, sometimes several generations. It is also often more about changes in attitudes, expectations and social norms than it is about physical things and material provisions. Donors may like to see quick and measurable results – and that their generosity pays off – but meaningful benefits can only be achieved through long-term commitment and reliable and respectful partnerships.
4. *Be positive*: Too often aid is seen as a way of filling local needs and providing what people lack. Whilst such resources can be valuable, it is better to think of the assets and capabilities societies and their people have first and see 'recipients' not as passive, helpless and lacking, but as people and agencies that are resilient, capable, and that contain potential. Working with and supporting – not replacing – those local assets is critical if aid is to achieve sustained improvements.
5. *Be collaborative*: Donors should not try to pursue their own interests ahead of others but instead act in concert with both recipient agencies and other donors so that substantial resources can be pooled and committed to activities that can have long-term and large-scale impact.
6. *Be transparent*: Everyone knows that aid is given, at least in part, out of self-interest whether that be with the expectation of some economic or political return or merely that it helps build relationships or makes us feel better about our wealthy societies. It also informs the public in donor countries in a way that is not patronising and more likely to be sustainable. It is better to be honest with regard to these donor motives than hide behind supposed singular altruism. Aid involves two-way reciprocal relationships. Acknowledging these openly helps build trust, understanding and transparency and allows for continual negotiation and accountability.

Conclusion

Aid has been part of the development landscape for the past 70 years and has involved billions of dollars flowing from donors to recipients and unquantifiable flows – both financial and political – in return. Ever since the Marshall Plan for the reconstruction of post-war Europe set the initial blueprint for aid in the late 1940s, aid has been used to accelerate development processes whilst also fulfilling the strategic and political goals of donors. Aid has always invoked a rationale based on 'development', albeit in many different forms and with various means, but it has also always been contingent on donors perceiving that it returned some benefits for them, whether in terms of economic or political gains, international standing and prestige, or satisfaction of domestic political demands.

Undoubtedly over the past 70 years much aid has been misguided, misdirected, stolen, lost or been harmful. We can agree that the allocation of aid has often not fulfilled its promises nor met its objectives. Despite the generations of aid and the massive volumes of financial disbursements, absolute poverty persists after all and fundamental global inequalities in wealth remain. We agree too that some forms of aid may well have exacerbated these levels of poverty and perpetuated social and economic inequalities. Aid has also been used, sometimes quite blatantly, as another way of imposing the will of rich and powerful on the countries and communities of the Global South. Critics, on the political left and right, are correct to keep scrutinising aid in terms of the motives of donors and recipients, and the impacts that aid has on all parties involved, whether open or covert. Aid – bad aid – can constrain, distort or limit the opportunities of individuals, groups and countries to improve their well-being.

However, many forms of aid do work. Aid can bring relief from disasters and hardship. Aid can provide needed resources and ideas. And, at its best, aid can help build the assets and capabilities of people and their governments so that they can pursue the development objectives and strategies that they define, manage and own. Whilst we might feel that aid is therefore justified because of this proven potential and has an underlying ethic of social justice and human concern, the reality of aid is that it can only be sustained at high levels (beyond the philanthropical support of millions of private individuals) if it has the backing of donor governments who can raise substantial and regular budgets to support major aid-funded development efforts. Aid therefore is predicated on an often-uncomfortable foundation of altruist narratives to 'sell' the idea and public motivation for aid alongside more opaque considerations of donor self-interest. Even though we may stress the value of the former in terms of a progressive transformation of the global economy, we have to accept that the latter will always need to be recognised and interrogated if aid is to continue.

Aid has also changed greatly over the past seven decades. We might like to suggest that it has become successively more effective over time as lessons are learned and good practices shared. However, sadly, what we have seen instead is evocative of a pendulum, where aid policies of donors have swung back and forth, sometimes stressing poverty alleviation, social justice and altruism, and at others highlighting economic growth and donor self-interest. Practices have also moved to and fro, so that what we see today are some elements of aid that have retrogressed to assumptions and modes that were in vogue 50 or 60 years ago.

It might be the case that aid, as we understand it today, may fade away and evolve into a quite different form of self-interested foreign and economic policy, not even carrying the epithet 'aid'. We certainly see the seeds of this in the evolving retroliberal restructuring of aid, cuts in aid budgets, the discourse of 'shared prosperity' and mechanisms such as Viability Gap Funding. Yet we retain a degree of optimism. Because aid has been such an important part of foreign policy for so long, it is unlikely that donors will dispose of it completely. It does build diplomatic and personal relationships, it does bring a return to donors and it does satisfy important public demands to address issues of global poverty, inequality and environmental degradation. We believe aid will exist for many years to come, but it is likely to continue to evolve, regress, reinvent itself, expand and contract, shift geographies and experiment with new and old methods. Despite these inevitable fluctuations and contested debates, aid will remain, as it has always been, a potentially critical force for good within the goal of reducing global poverty and inequality.

Summary

- There is a wide range of opinions concerning the definition, rationale, measurement and impact of aid.

- In terms of the impacts there are critics from both the political left and right. Broadly speaking, the left sees aid as a tool of imperialism which creates dependency and exacerbates the North–South divide and the right sees aid as something which distorts free-markets, undermines self-responsibility and leads to corruption.
- The rhetorical arguments of the right and left are seldom based on empirical studies. The evidence suggests that some aid has failed and some has succeeded.
- The aid landscape is shifting and currently there is a clear move towards retroliberal, private sector-based models that place donor self-interest much more firmly in the centre.
- It would be naïve to suggest that self-interest should not determine aid policy, for it always has. A sustainable future for aid requires that there be benefits for donors and recipients.
- It is possible to envisage three broad futures for aid – pessimistic, realistic and optimistic. We tend towards the optimistic and see the SDGs as a framework for the projection of a positive model of aid that will bring more costs than benefits.
- ODA is shifting in its nature and the range of aid types that exist is broadening. We feel we are moving into a 'post-ODA' world.
- We suggest six lessons for effective aid. If thoughtfully conceptualised, designed, applied, measured and evaluated aid remains a progressive force for global prosperity, sustainability, and justice.

Discussion questions

- Summarise the viewpoint from the left and the right of politics with respect to the impacts of aid on recipient and donor countries.
- What are the three views on the future of aid? Outline the global conditions that have given rise to each scenario and say which you think best predicts the future.
- What is meant by the relationship value of aid?
- Outline the six lessons for effective aid. How can these be criticised?
- Discuss the difference between a 'post-aid' and a 'post-ODA' world. Which is most useful in terms of achieving development progress in your opinion?

Websites

- Oxfam: 21st Century Aid, https://www.oxfam.org/en/research/21st-century-aid

Further reading

Easterly, W. (ed.) (2008) *Reinventing Foreign Aid.* MIT Press, London and Cambridge, MA.

Janus, H., Klingebiel, S. and Paulo, S. (2015) 'Beyond aid: A conceptual perspective of the transformation of development cooperation', *Journal of International Development* 27(2), 155–169.

Mawdsley, E., Savage, L. and Kim, S.M. (2014) 'A "post-aid world"? Paradigm shift in foreign aid and development cooperation at the 2011 Busan High Level Forum', *The Geographical Journal* 180(1), 27–38.

Mawdsley, E. (2019) 'South–South Cooperation 3.0? Managing the consequences of success in the decade ahead', *Oxford Development Studies* 47(3), 259–274.

Oxfam (2010) *21st Century Aid: Recognising Success and Tackling Failure.* Oxfam Briefing Paper 137.

References

Abdenur, A.E. (2015) 'Organisation and politics in South–South Cooperation: Brazil's technical cooperation in Africa', *Global Society* 29(3) 321–338.

Abrahams, J. (2017) 'Europe's risky experiment: Can aid be used to deter migration?', Devex https://www.devex.com/news/europe-s-risky-experiment-can-aid-be-used-to-deter-migration-90426?utm_source=website&utm_medium=box&utm_campaign=linking_strategy (accessed 12 September 2019).

Action Aid (2008) *Making Aid Accountable and Effective*. Action Aid, Johannesburg.

Addison, T., Morrissey, O. and Tarp, F. (2017) 'The macroeconomics of aid: Overview', *The Journal of Development Studies* 53(7) 987–997.

African Review (2019) 'More than 2,300 cataract surgeries with Cameroon Cataract Bond', *African Review*, http://www.africanreview.com/manufacturing/industry/more-than-2-300-cataract-surgeries-with-cameroon-cataract-bond (accessed 8 September 2019).

Agg, C. (2006) *Trends in Government Support for Non-Governmental Organizations: Is the "Golden Age" of the NGO Behind Us?* UNRISD, Geneva.

Ahmed, E. (2019) 'The politics of aid in Afghanistan', Observer Research Foundation, https://www.orfonline.org/expert-speak/the-politics-of-aid-in-afghanistan-52668/ (accessed 29 September 2019).

Alesina, A. and Dollar, D. (2002) 'Who gives foreign aid to whom and why?', *Journal of Economic Growth* 5 33–63.

Alfini, N. and Chambers, R. (2007) 'Words count: Taking a count of the changing language of British aid', *Development in Practice* 17(4–5) 492–504.

Anderson, M. (2015a) 'DfID to pump £735m into investment arm for private sector projects', *The Guardian*, https://www.theguardian.com/global-development/2015/jul/17/department-for-international-development-cdc-group-735m-uk-aid-private-sector (accessed 6 April 2018).

Anderson, M. (2015b) 'UK aid watchdog criticises DfID over partnerships with private sector', *The Guardian*, https://www.theguardian.com/global-development/2015/may/21/uk-aid-watchdog-dfid-public-private-partnerships-icai (accessed 6 April 2018).

Arndt, C., Jones, S. and Tarp, F. (2015) 'Assessing foreign aid's long-run contribution to growth and development', *World Development* 69 6–18.

Asante, A.D., Martins, N., Otim, M.E. and Dewdney, J. (2014) 'Retaining doctors in rural Timor-Leste: A critical appraisal of the opportunities and challenges', *Bulletin of the World Health Organization* 92(4) 277–282.

Asante, A.D., Negin, J., Hall, J., Dewdney, J. and Zwi, A. B. (2012) 'Analysis of policy implications and challenges of the Cuban health assistance program related to human resources for health in the Pacific', *Human Resources for Health* 10, 10.

Baker, T., Evans, J. and Hennigan, B. (2019) 'Investable poverty: Social investment states and the geographies of poverty management', *Progress in Human Geography*, online first 1–21 doi:10.1177/0309132519849288.

Banks, G., Murray, W.E., Overton, J. and Scheyvens, R. (2012) 'Paddling on one side of the canoe? The changing nature of New Zealand's development assistance programme', *Development Policy Review* 30(2) 169–186.

Barcelos, P. and De Angelis, G. (eds.) (2016) *International Development and Human Aid: Principles, Norms and Institutions for the Global Sphere*. Edinburgh University Press, Edinburgh.

Bauer, P. (1976) *Dissent on Development*. Harvard University Press, Cambridge Mass.

Bellù, L.G. (2011) 'Development and development paradigms: A (reasoned) review of prevailing visions', *ESYPol Module* 102. FAO, Rome.

Berndt, C. and Wirth, M. (2018) 'Market, metrics, morals: The Social Impact Bond as an emerging social policy instrument', *Geoforum* 90 27–35.

Bertram, I.G. (2004) 'On the convergence of small island economies with their metropolitan patrons', *World Development* 32(2) 343–364.

Bertram, I.G. (2006) 'The MIRAB model in the twenty-first century', *Asia Pacific Viewpoint* 47(1) 1–13.

Bertram, G. (2018) 'Why does the Cook Islands still need overseas aid?', *Journal of Pacific History* 53(1) 44–63.

Brant, P. (2016) 'Mapping Chinese aid in the Pacific' in Powles, M. (ed) *China and the Pacific: The View from Oceania*, Victoria University Press, Wellington, 173–175.

Booth, D. (2011) 'Aid, institutions and governance: what have we learned?', *Development Policy Review* 29(1) 5–26.

Bräutigam, D. (2011) 'Aid "with Chinese characteristics": Chinese foreign aid and development finance meet the OECD/DAC aid regime', *Journal of International Development* 23(5) 752–764.

Briggs, R.C. (2014) 'Aiding and abetting: Project aid and ethnic politics in Kenya', *World Development* 64 194–205.

Briggs, R.C. (2018) 'Poor targeting: A gridded spatial analysis of the degree to which aid reaches the poor in Africa', *World Development* 103 133–148.

Brockington, D. (2014) *Celebrity Advocacy and International Development*. Routledge, London.

Brown, S. and Grävingholt, J. (eds) (2016) *The Securitization of Foreign Aid*. Palgrave Macmillan, London.

Buiter, W.H. (2007) "Country ownership': A term whose time has gone', *Development in Practice* 17(4–5) 647–652.

Burnside, C. and Dollar, D. (2000) 'Aid, policies and growth', *American Economic Review* 90(4) 847–869.

Cassen, R. (1994) *Does Aid Work?* (2nd ed) Oxford University Press, New York.

Chambers, R. (1983) *Rural Development: Putting the Last First*. Longman, Harlow.

Chambers, R. (2004) 'Ideas for development: reflecting forwards', *IDS Working Paper* 238 40pp.

Cheney, C. (2017) 'German foreign aid is at a record level and rising. Here is how it works', Devex, https://www.devex.com/news/german-foreign-aid-is-at-a-record-high-and-rising-here-is-how-it-works-89366?utm_source=website&utm_medium=box&utm_campaign=linking_strategy (accessed 12 September 2019).

Choudry, A. and Kapoor, D. (eds) (2013) *NGOization: Complicity, Contradictions and Prospects*. Zed Books, London.

Christie, R. (2012) 'The pacification of soldiering, and the militarization of development: Contradictions inherent in provincial reconstruction in Afghanistan', *Globalizations* 9(1) 53–71.

Collier, P. (2007) *The Bottom Billion: Why the Poorest Countries are Failing and What Can be Done About it*. Oxford University Press, London and New York.

Collier, P. (2016) 'The ethical foundations of aid: Two duties of rescue' *BSG Working Paper Series* BSG-WP-2016/016, Blavatnik School of Government, University of Oxford.

Collier, P. and Dollar, D. (2002) 'Aid allocation and poverty reduction', *European Economic Review* 46(8) 1475–1500.

Cowen, M. (1984) 'Early years of the Colonial Development Corporation: British state enterprise overseas during late colonialism', *African Affairs* 83(330) 63–75.

Craig, D. and Porter, D. (2003) 'Poverty reduction strategy papers: A new convergence', *World Development* 31(1) 53–69.

Craig, D. and Porter, D. (2006) *Development Beyond Neoliberalism? Governance, Poverty Reduction and Political Economy*. Routledge, London and New York.

Culp, J. (2016) 'Toward another kind of development practice' in Barcelos, P., and De Angelis, G. (eds.) *International Development and Human Aid: Principles, Norms and Institutions for the Global Sphere*. Edinburgh University Press, Edinburgh, pp.79–107.

DAC (2006) *DAC in Dates: The History of OECD's Development Assistance Committee*. OECD, Paris.

DAC Secretariat (2016a) *The Scope and Nature of 2016 HLM Decisions Regarding the ODA-Eligibility of Peace and Security-Related Expenditures*, Paris, OECD-DAC, http://www.oecd.org/dac/HLM_ODAeligibilityPS.pdf (accessed 15 May 2017).

DAC Secretariat (2016b) *ODA Reporting of In-Donor Country Refugee Costs. Members' Methodologies for Calculating Costs*, Paris, OECD-DAC, https://www.oecd.org/dac/stats/RefugeeCostsMethodologicalNote.pdf (accessed 15 May 2017).

Dalgaard, C.-J., and Hansen, H. (2017) 'The return to foreign aid', *The Journal of Development Studies* 53(7) 998–1018.

de Haan, A. (2009) *How the Aid Industry Works: An Introduction to International Development*. Kumarian Press, Sterling.

Denizer, C., Kaufmann, D. and Kraay, A. (2013) 'Good countries or good projects? Macro and micro correlates of World Bank project performance', *Journal of Development Economics* 105 288–302.

Devex (2013) 'Callan, Blak and Thomas on China's foreign aid and investment', https://www.devex.com/news/sponsored/callan-blak-and-thomason-china-s-foreign-aid-and-investment-80594 (accessed 11 September 2019).

Devex (2016) 'ODA redefined: What you need to know', https://www.devex.com/news/oda-redefined-what-you-need-to-know-87776 (accessed 11 August 2017).

DFAT (n.d.) 'About the New Colombo Plan', DFAT Australia, https://www.dfat.gov.au/people-to-people/new-colombo-plan/about/Pages/about (accessed 3 March 2018).

DFID (2011a) 'DFID's approach to value for money (VfM)', Department for International Development, https://assets.publishing.service.gov.uk/government/uploads/system/uploads/attachment_data/file/49551/DFID-approach-value-money.pdf (accessed 6 October 2017).

DFID (2011b) 'The engine of development: The private sector and prosperity for poor people', Department for International Development, https://www.gov.uk/government/publications/the-engine-of-developmentthe-private-sector-and-prosperity-for-poor-people (accessed 4 October 2017).

DFID (2017) 'Business case summary sheet: Invest Africa', iati.dfid.gov.uk>iati_documents/24923374.odt (accessed 18 April 2019).

Dijkstra, G. (2018) 'Aid and good governance: Examining aggregate unintended effects of aid', *Evaluation and Program Planning* 68 225–232.

Dollar, D. and Pritchett, L. (1998) *Assessing Aid: What Works, What Doesn't, and Why*. Oxford University Press, Washington DC.

Dominiczak, P. (2016) 'Aid department "failed" taxpayers over St Helena Airport that can't take planes because of wind', *The Telegraph* https://www.telegraph.co.uk/news/2016/12/14/aid-department-failed-taxpayers-st-helena-airport-cant-take/ (accessed 4 September 2017).

Dunning, T. (2004) 'Conditioning the effects of aid: Cold War politics, donor credibility, and democracy in Africa', *International Organization* 58(2) 409–423.

Easterly, W. (2003) 'Can foreign aid buy growth?', *Journal of Economic Perspectives* 17(3) 23–48.

Easterly, W. (2005) 'What did structural adjustment adjust? The association of policies and growth with repeated IMF and World Bank adjustment loads', *Journal of Development Economics* 76 1S, 1–22.

Easterly, W. (2006) *The White Man's Burden: Why the West's Efforts to Aid the Rest Have Done So Much Ill and So Little Good*. Oxford University Press, New York and Oxford.

Easterly, W. (2007) 'Was development assistance a mistake?', *American Economic Review Papers and Proceedings* 97(2) 328–332.

Easterly, W. (ed.) (2008) *Reinventing Foreign Aid*. MIT Press, London and Cambridge, Mass.

Easterly, W. and Pfutze, T. (2008) 'Where does the money go? Best and worst practices in foreign aid', *Journal of Economic Perspectives* 22(2) 29–52.

ECLAC (1990) *Changing Production Patterns with Social Equity*. Economic Commission for Latin America and the Caribbean, Santiago.

Edwards, J. (2019) 'New Zealand aid and dairy development in Sri Lanka', unpublished Master of Development Studies thesis, Victoria University of Wellington, Wellington.

Erickson, A. (2017) 'America sanctioned Venezuela. Then it offered $5 million in aid to Harvey victims', *The Washington Post* https://www.washingtonpost.com/news/worldviews/wp/2017/08/30/venezuela-enemy-of-the-u-s-offered-5-million-in-aid-to-harvey-victims/?utm_term=.d0eece714420 (accessed 6 September 2017).

Escobar, A. (1995) *Encountering Development: The Making and Unmaking of the Third World*. Princeton University Press, Princeton.

Eurodad (2008) *Outcome-Based Conditionality: Too Good to be True?* European Network on Debt and Development. Eurodad, Brussels.

European Commission (2018) *Budget Support: Trends and Results 2018*. Directorate-General, International Cooperation and Development, European Commission, Luxembourg.

Eyben, R. (2007) 'Harmonisation: how is the orchestra conducted?', *Development in Practice* 17(4–5) 640–646.

Eyben, R. (2013) 'Struggles in Paris: The DAC and the purposes of development aid', *European Journal of Development Research* 25(1) 78–91.

Eyben, R. and Savage, L. (2013) 'Emerging and submerging powers: Imagined geographies in the new development partnership at the Busan high fourth level forum', *The Journal of Development Studies* 49(4) 457–469.

Fayez, H. (2012) 'The role of foreign aid in Afghanistan's reconstruction: A critical assessment', *Economic and Political Weekly* 47(39) 65–70.

Fejerskov, A.M. (2015) 'From unconventional to ordinary? The Bill and Melinda Gates Foundation and the homogenizing effects of International Development Cooperation', *Journal of International Development* 27(7) 1098–1112.

Fraser, A., Tan, S., Lagarde, M. and Mays, N. (2018) 'Narratives of promise, narratives of caution: A review of the literature on social impact bonds', *Social Policy and Administration* 52(1) 4–28.

Freire, P. (1970) *Pedagogy of the Oppressed*. Herder and Herder, New York.

Fuchs, A. and Vadlamannati, K.C. (2013) 'The needy donor: An empirical analysis of India's aid motives', *World Development* 44, 110–128.

Galiani, S., Knack, S., Xu, L.C. and Zou, B. (2017) 'The effect of aid on growth: Evidence from a quasi-experiment', *Journal of Economic Growth* 22(1) 1–33.

Gamlen, A. (2014) 'The new migration and development pessimism', *Progress in Human Geography* 38(4) 581–597.

Gibson, C.C., Hoffman, B.D. and Jablonski, R.S. (2015) 'Did aid promote democracy in Africa? The role of technical assistance in Africa's transitions', *World Development* 68 323–335.

Gibson, C.C., Andersson, K., Ostrom, E. and Shivakumar, S. (2005) *The Samaritan's Dilemma: The Political Economy of Development Aid*. Oxford University Press, Oxford.

Gibson, J. and Mckenzie, D. (2012) 'The economic consequences of "brain drain" of the best and brightest: microeconomic evidence from five countries', *The Economic Journal* 122(560), 339–375.

GIZ (n.d.) 'Chile' https://www.giz.de/en/worldwide/388.html (accessed 9 September 2019).

Glassman, A. and Oroxom, R. (2017) 'Another one joins the DIB: OPIC commits $2 million to a development impact bond on cataract surgery', Center for Global Development, https://www.cgdev.org/blog/another-one-joins-dib-opic-commits-2-million-development-impact-bond-cataract-surgery (accessed 6 April 2018).

Glennie, J. (2008) *The Trouble with Aid: Why Less Could Mean More for Africa*. Zed Books, London.

Glennie, J. (2011) 'Yes, the Paris declaration on aid has problems but it's still the best we have', *The Guardian*, https://www.theguardian.com/global-development/poverty-matters/2011/nov/18/paris-declaration-aid-effectiveness-necessary (accessed 16 August 2019).

Gould, J. (2005) *The New Conditionality: The Politics of Poverty Reduction Strategies*. Zed Books, London.

Government of Denmark (2013) *Denmark's Integrated Stabilisation Engagement in Fragile and Conflict-Affected Areas of the World*. Ministry of Foreign Affairs, Ministry of Defence and Ministry of Justice, Copenhagen.

Greenhill, R., Prizzon, A. and Rogerson, A. (2013) 'The age of choice: developing countries in the new aid landscape', *ODI Working Papers* no. 364, ODI, London.

Griffin, K. (1991) 'Foreign aid after the Cold War', *Development and Change* 22(4) 645–648.

Gronemeyer, M. (2010) 'Helping' in W. Sachs (ed) *The Development Dictionary* 2nd ed. Zed Books, London and New York, pp. 55–73.

Gulrajani, N. (2017) 'Bilateral donors and the age of the national interest: What prospects for challenge by development agencies?', *World Development* 96 375–389.

Gulrajani, N. and Calleja, R. (2019) The Principled Aid Index. *ODI Policy Briefing.* ODI, London.

Gulrajani, N. and Faure, R. (2019) 'Donors in transition and the future of development cooperation: What do the data from Brazil; India, China, and South Africa reveal?', *Public Administration and Development* 1–14 early view online doi: 10.1002/pad.1861.

Gulrajani, N. and Swiss, L. (2019) 'Donor proliferation to what ends? New donor countries and the search for legitimacy', *Canadian Journal of Development Studies / Revue Canadienne d'études du développement* 40(3) 348–368.

Gutiérrez, A. and Jaimovich, D. (2017) 'A new player in the international development community? Chile as an emerging donor', *Development Policy Review* 35(6) 839–858.

Gwynne R.N., Klak, T. and Shaw, D.J.B. (2014) *Alternative Capitalisms: Geographies of Emerging Regions*. Routledge, London.

Hadley, S. and Miller, M. (2016) 'PEFA: What is it good for?', *ODI Discussion Paper April 2016*. Overseas Development Institute, London.

Hanson, F. (2008) 'The dragon looks South', *Analysis Paper*, Lowy Institute for International Policy, Sydney.

Hanson, F. (2009) 'China: Stumbling through the Pacific', *Policy Brief*, Lowy Institute for International Policy, Sydney.

Hardin, G. (1974) 'Living on a lifeboat', *Bioscience* 24(10) 561–568.

Harman, S. and Williams, D. (2014) 'International development in transition', *International Affairs* 90(4) 925–941.

Hart, D. (2010) 'D/developments after the meltdown', *Antipode* 41(s1) 117–141.

Harvey, D. (2005) *A Brief History of Neoliberalism.* Oxford University Press, New York.

Hayter, T. (1971) *Aid as Imperialism.* Penguin Books, Harmondsworth.

Heinrich, T., Machain, C.M. and Oestman, J. (2017) 'Does counterterrorism militarize foreign aid? Evidence from sub-Saharan Africa', *Journal of Peace Research* 54(4) 527–541.

Hickey, S. (2013) 'Beyond "poverty reduction through good governance": The new political economy of development in Africa', *New Political Economy* 17(5) 683–690.

Hintjens, H. and Hodge, D. (2012) 'The UK Caribbean overseas territories: Governing unruliness amidst the extra-territorial EU', *Commonwealth & Comparative Politics* 50(2) 190–225.

Hirschman, A.O. (1958) *The Strategy of Economic Development.* Yale University Press, New Haven and London.

Howard, P. and Rothenberg, P.J. (1993) 'Foreign aid and the question of fungibility', *Review of Economics and Statistics* 75(2) 258–265.

Howell, J. and Lind, J. (2008) 'Changing donor policy and practice in civil society in the post-9/11 aid context', *Third World Quarterly* 30(7) 1279–1296.

Hughes, H. (2003) *Aid has Failed the Pacific,* Issue Analysis no. 33, The Centre for Independent Studies, Sydney.

Hyden, G. (2008) 'After the Paris declaration: Taking on the issue of power', *Development Policy Review* 26(3) 259–274.
ICAI (2015) 'Business in development', Report 43, Independent Commission for Aid Impact, London.
ICAI (2019a) 'CDC's investments in low-income and fragile states: A performance review', Independent Commission for Aid Impact, London.
ICAI (2019b) 'The current state of UK aid: A synthesis of ICAI findings from 2015 to 2019', Independent Commission for Aid Impact, London.
IISD (n.d.) 'Private Infrastructure Development Group', https://iisd.org/credit-enhancement-instruments/institution/private-infrastructure-development-group/ (accessed 22 October 2019).
International Poverty Centre (2007) *Does Aid Work? – For the MDGs*. Poverty in Focus International Poverty Centre, Brasilia.
Janus, H., Klingebiel, S. and Paulo, S. (2014) '"Beyond aid" and the future of development cooperation', German Development Institute *Briefing Paper* 6/2014), Deutsches Institut für Entwicklungspolitik (DIE), Bonn.
Janus, H., S. Klingebiel and Paulo, S. (2015) 'Beyond aid: A conceptual perspective of the transformation of development cooperation', *Journal of International Development* 27(2) 155–169.
Jensen, E.H. (2019) 'The Peace and Stabilisation Fund (PSF)' Presentation to the 13th Seoul ODA International Conference 'Promoting Co-Prosperity to Achieve the SDGs' 19 September 2019.
Jomo, K.S. and Chowdhury, A. (2019) 'World Bank dispossessing rural poor', Inter Press Service, http://www.ipsnews.net/2019/04/world-bank-dispossessing-rural-poor/ (accessed 18 April 2019).
Jones, S. and Tarp, F. (2016) 'Does foreign aid harm political institutions?', *Journal of Development Economics* 118 266–281.
Kay. C. (2011) *Latin American Theories of Development and Underdevelopment*. Routledge, London.
Kersting, E. and Kilby, C. (2014) 'Aid and democracy redux', *European Economic Review* 67 125–143.
Kilby, P. (2018) 'DAC is dead? Implications for teaching development studies', *Asia Pacific Viewpoint* 59(2) 226–234.
Kim, S. and Lightfoot, S. (2011) 'Does "DAC-ability" really matter? The emergence of non-DAC donors: Introduction to policy arena', *Journal of International Development* 23(5) 711–721.
Knack, S., Rogers, F.H. and Eubank, N. (2011) 'Aid quality and donor rankings', *World Development* 39(11) 1907–1917.
Koeberle, S. and Stavreski, Z. (2006) 'Budget support: Concepts and issues', in Koeberle, S., Stavreski, Z. and Walliser, J., (eds) *Budget Support as More Effective Aid*. World Bank. Washington DC, pp.3–27.
Koeberle, S., Stavreski, Z. and Walliser, J. (eds) (2006) *Budget Support as More Effective Aid*. World Bank, Washington DC.
Kotsadam, A., Østby, G., Rustad, S.A., Tollefsen, A.F., and Urdal, H. (2018) 'Development aid and infant mortality. Micro-level evidence from Nigeria', *World Development* 105 59–69.
Kramer, W.M. (2007) 'Corruption and fraud in international aid projects', *U4 Brief*, 2007 no.4, CMI, Bergen.

Kumar, R. (2019) *The Business of Changing the World: How Billionaires, Tech Disrupters, and Social Entrepreneurs are Transforming the Global Aid Industry*. Beacon Press, Boston.

Kunz, D.B. (1997) 'The Marshall Plan reconsidered: A complex of motives', *Foreign Affairs*, 76(3) 162–170.

Lancaster, C. (2007) *Foreign Aid: Diplomacy, Development, Domestic Politics*. University of Chicago Press, Chicago.

Lei Ravelo, J. (2017) 'OECD aid reaches record high but more money is spent domestically', Devex https://www.devex.com/news/oecd-aid-reaches-record-high-but-more-money-is-spent-domestically-90034?utm source=website&utm medium=box&utm campaign=linking strategy (accessed 12 September 2019).

Leiva, F. (2008) 'Toward a critique of Latin American neostructuralism', *Latin American Politics and Society* 50 1–25.

Levine, R. (2007) *Case Studies in Global Health: Millions Saved*. Jones and Bartlett Publishers, Sudbury.

Levy, B. (2014) *Working with the Grain: Integrating Governance and Growth in Development Strategies*. Oxford University Press, New York.

Lewis, O. (2017) 'RSE scheme turns 10: The crucial role seasonal workers play in New Zealand's growing wine industry', https://www.stuff.co.nz/business/farming/agribusiness/94490362/rse-scheme-turns-10-the-crucial-role-seasonal-workers-play-in-new-zealands-growing-wine-industry (accessed 6 April 2018).

Lewis, W.A. (1954) 'Economic development with unlimited supplies of labour', *The Manchester School* 22(2) 139–191.

Loraque, J. (2018) 'Development Impact Bonds: Bringing innovation to education development financing and delivery', *Childhood Education* 94(4) 64–68.

Martens, B., Mummert, U., Murrell,.P, Seabright, P. and Ostrom, E. (2002) *The Institutional Economics of Foreign Aid*. Cambridge University Press, Cambridge.

Masoud, M.R., Kalliapan, S.R., Ismail, N.W. and Azman-Saini W.N.W. (2015) 'Effect of foreign aid on corruption: Evidence from Sub-Saharan African countries', *International Journal of Social Economics* 42(1) 47–63.

Mawdsley, E. (2010) 'The non-DAC donors and the changing landscape of foreign aid: the (in)significance of India's development cooperation with Kenya', *Journal of Eastern African Studies* 4(2) 361–379.

Mawdsley, E. (2012a) *From Recipients to Donors: Emerging Powers and the Changing Development Landscape*. Zed Books, London.

Mawdsley, E. (2012b) 'The changing geographies of foreign aid and development cooperation: Contributions from gift theory', *Transactions of the Institute of British Geographers* 256–272.

Mawdsley, E. (2012c) 'The changing geographies of foreign aid and development cooperation: Contributions from gift theory', *Transactions of the Institute of British Geographers* 37(2) 256–272.

Mawdsley, E. (2014) 'Human rights and South-South development cooperation: Reflections on the "rising powers" as international development actors', *Human Rights Quarterly* 36(3) 630–652.

Mawdsley, E. (2015) 'DFID, the private sector, and the re-centring of an economic growth agenda in international development', *Global Society* 29(3) 339–358.

Mawdsley, E. (2017) 'Development Geography 1: Cooperation, Competition and Convergence between "North" and "South"', *Progress in Human Geography* 41(1) 108–117.

Mawdsley, E. (2018) 'The 'Southernisation' of development?', *Asia Pacific Viewpoint* 59(2), 173–185.

Mawdsley, E. (2019) 'South–South Cooperation 3.0? Managing the consequences of success in the decade ahead', *Oxford Development Studies* 47(3) 259–274.

Mawdsley, E., Murray, W.E., Overton, J, Scheyvens, R. and Banks, G.A. (2018) 'Exporting stimulus and 'shared prosperity': Re-inventing aid for a retroliberal era', *Development Policy Review* 36 O25–O43.

Mawdsley, E., Savage, L. and Kim, S.M. (2014) 'A "post-aid world"? Paradigm shift in foreign aid and development cooperation at the 2011 Busan High Level Forum', *The Geographical Journal* 180(1) 27–38.

MBIE (Ministry of Business, Innovation, and Employment New Zealand) (2016) 'The Remittance Pilot Project The economic benefits of the Recognised Seasonal Employer work policy and its role in assisting development in Samoa and Tonga – final report', https://www.immigration.govt.nz/documents/statistics/remittance-pilot-project-final-report.pdf (accessed 6 April 2018).

McEwan, C. and Mawdsley, E. (2012) 'Trilateral development cooperation: Power and politics in emerging aid relationships', *Development and Change* 43(6) 1185–1209.

McGregor, A., Challies, E., Overton, J. and Sentes, L. (2013) 'Developmentalities and donor−NGO relations: Contesting foreign aid policies in New Zealand/ Aotearoa', *Antipode* 45(5) 1232–1253.

McMichael, P. (2017) *Development and Social Change* (6th ed). Pine Forge, Thousand Oaks.

McVeigh, K. (2017) 'Ed Sheeran Comic Relief film branded 'poverty porn' by aid watchdog', *The Guardian*, https://www.theguardian.com/global-development/2017/dec/04/ed-sheeran-comic-relief-film-poverty-porn-aid-watchdog-tom-hardy-eddie-redmayne).

Mekasha, T.J. and Tarp, F. (2019) 'A meta-analysis of aid effectiveness: Revisiting the evidence', *Politics and Governance* 7(2) 5–28.

Metzger, L., Nunnenkamp, P. and Mahmoud, T.O. (2010) 'Is corporate aid targeted to poor and deserving countries? A case study of Nestlé's aid allocation', *World Development* 38(3) 228–243.

MFAT (2011) 'Value for money guideline', MFAT, Wellington, https://www.mfat.govt.nz/en/aid-and-development/working-with-us/tools-and-guides-for-aid-activities/ (accessed 4 July 2018).

Ministry of Foreign Affairs of Denmark (2014) 'Evaluation of the Danish Peace and Stabilisation Fund', Evaluation Department, Ministry of Foreign Affairs of Denmark, Copenhagen.

Minoiu, C. and Reddy, S.G. (2009) *Development Aid and Economic Growth: A Positive Long-Run Relation.* IMF Working Paper WP/09/118, IMF, Washington.

Molenaers, N., Dellepiane, S. and Faust, J. (2015) 'Political conditionality and foreign aid', *World Development* 75 2–12.

Morrissey, O. (2001) 'Does aid increase growth?', *Progress in Development Studies* 1(1) 37–50.

Moyo, D. (2010) *Dead Aid: Why Aid is Not Working and How There is Another Way for Africa*. Farrar, Straus and Giroux, New York.

Murray, W. E. and Overton, J. (2011a) 'Neoliberalism is dead, long live neoliberalism. Neostructuralism and the new international aid regime of the 2000s', *Progress in Development Studies* 11(4) 307–319.

Murray, W.E. and Overton, J. (2011b) 'The inverse sovereignty effect: Aid, scale and neostructuralism in Oceania', *Asia Pacific Viewpoint* 52(3) 272–284.

Murray, W.E. and Overton, J. (2016) 'Retroliberalism and the new aid regime of the 2010s', *Progress in Development Studies*, 16(3) 1–17.

Nussbaum, M. (2000) *Women and Human Development: The Capabilities Approach*. Cambridge University Press, Cambridge.

Oakman, D. (2000) 'The seed of freedom: Regional security and the Colombo Plan', *Australian Journal of Politics and History* 46(1) 67–85.

Oakman, D. (2010) *Facing Asia: A History of the Colombo Plan*. ANU E Press, Canberra.

Ocampo, J. (1993) 'Terms of trade and center-periphery relations', in O. Sunkel (ed.), *Development from Within: Toward a Neostructuralist Approach for Latin America*. Lynne Rienner Publishers, Boulder, pp. 333–360.

OECD (2008) *The Paris Declaration on Aid Effectiveness and the Accra Agenda for Action*. Paris, OECD, http://www.oecd.org/dac/effectiveness/34428351.pdf (accessed 20 February 2014).

OECD (2016) 'In-donor country refugee costs reported as ODA by OECD-DAC members', http://www.oecd.org/dac/financing-sustainable-development/In-donor-refugee-costs-in-ODA.pdf (accessed 16 December 2016).

OECD (2017) 'Resource receipts' https://public.tableau.com/views/Non-ODAflows/ResourceReceipts?:embed=y&:display_count=no?&:showVizHome=no#1 (accessed 24 March 2019).

OECD (2019a) 'DAC List of ODA Recipients: Effective for reporting on 2018, 2019 and 2020 flows', https://www.oecd.org/dac/financing-sustainable-development/development-finance-standards/DAC_List_ODA_Recipients2018to2020_flows_En.pdf (accessed 29 March 2019).

OECD (2019b) 'What is ODA?', Development Co-operation Directorate, April 2019. https://www.oecd.org/dac/stats/What-is-ODA.pdf (accessed 4 October 2019).

OECD (n.d.) 'Official development assistance – definition and coverage', http://www.oecd.org/dac/stats/officialdevelopmentassistancedefinitionandcoverage.htm#Definition (accessed 4 October 2019).

OPIC (2017) 'Cameroon cataract development impact loan offers innovative approach to prevent blindness', https://www.opic.gov/press-releases/2017/cameroon-cataract-development-impact-loan-offers-innovative-approach-prevent-blindness (accessed 6 April 2018).

Orlina, E.C. (2017) 'Top USAID contractors for 2016', DevEx, https://www.devex.com/news/top-usaid-contractors-for-2016-90202 (accessed 5 September 2017).

Overton, J. and Murray, W.E. (2018) 'Aid and sovereignty: Neostructuralism, retroliberalism, and the recasting of relationships in Oceania', in Brinklow, L. and Sinclair, J. (eds) *Islands Economic Cooperation Forum: Annual Report on Global Islands 2017*. Island Studies Press, Charlottetown PEI, pp. 163–186.

Overton, J., Murray W.E. and McGregor A. (2013) 'Geographies of aid: a critical research agenda', *Geography Compass* 7(2) 116–127.

Overton, J., Murray, W.E., Prinsen, G., Ulu, A. and Wrighton, N. (2019) *Aid, Ownership and Development: The Inverse Sovereignty Effect in the Pacific Islands*. Routledge, London.

Overton, J., Prinsen, G., Murray, W.E. and Wrighton, N. (2012) 'Reversing the tide of aid: Investigating development policy sovereignty in the Pacific', *Journal de la Société des Océanistes* 135 229–242.

Oxfam (2010) *21st Century Aid: Recognising Success and Tackling Failure*. Oxfam Briefing Paper 137.

Peck, J. (2010) 'Zombie neoliberalism and the ambidextrous state', *Theoretical Criminology* 14(1) 104–110.

Peck, J., Theodore, N. and Brenner, N. (2010) 'Postneoliberalism and its malcontents', *Antipode* 41(s1) 94–116.

PEFA (n.d.) 'What is PEFA?', https://pefa.org/about (accessed 3 September 2019).

Perrone, A. and Healy, H. (2018) 'Humanitarian action: The facts', *New Internationalist* issue 511 16–17.

Prebisch, R. (1962) 'The economic development of Latin America and its principal problems', *Economic Bulletin for Latin America* 7(1) 1–22.

Price, R.A. (2019) 'ODA and non-ODA resources in combination to address violent conflict', *K4D Helpdesk Report* 566, Institute of Development Studies, Brighton.

Prinsen, G., Lafoy, Y. and Migozzi, J. (2017) 'Showcasing the sovereignty of non-self-governing islands: New Caledonia', *Asia Pacific Viewpoint* 58(3) 331–346.

Provost, C. (2014) 'British aid money invested in gated communities and shopping centres', *The Guardian*, https://www.theguardian.com/global-development/2014/may/02/british-aid-money-gated-communities-shopping-centres-cdc-poverty (accessed 8 April 2018).

Pugh, J., Gabay, C. and Williams, A. (2013) 'Beyond the securitisation of development: The limits of intervention, developmentalisation of security and repositioning of purpose in the UK Coalition government's policy agenda', *Geoforum* 44 193–201.

Radelet, S. and Levine, R. (2008) 'Can we build a better mousetrap? Three new institutions designed to improve aid effectiveness' in W. Easterly (ed.) *Reinventing Foreign Aid*. MIT Press, London and Cambridge Mass., pp. 431–460.

Reinert, E. and Jomo, K.S. (2008) 'The Marshall Plan at 60: The General's successful war on poverty', *UN Chronicle* archived at https://web.archive.org/web/20090419195002/http://www.un.org/Pubs/chronicle/2008/webarticles/080103_marshallplan.html.

Reinikka, R. (2008) 'Donors and service delivery' in W. Easterly (ed.) *Reinventing Foreign Aid*. MIT Press, London and Cambridge Mass., pp. 179–199.

Richey, L.A. and Ponte, S. (2011) *Brand Aid: Shopping Well to save the World*. University of Minnesota Press, Minneapolis.

Riddell, R.C. (2007) *Does Foreign Aid Really Work?* Oxford University Press, New York.

Rimmer, D. (2000) 'Aid and corruption', *African Review* 91 121–128.

Rist, G. (1997) *The History of Development*. London, Zed Books.

Roberts, S.M. (2014) 'Development capital: USAID and the rise of development contractors', *Annals Association of American Geographers* 104(5) 1030–1051.

Rocha Menocal, A. and Denney, L. with Geddes, M. (2011) *Informing the Future of Japan's ODA. Part one: Locating Japan's ODA Within a Crowded and Shifting Marketplace.* Overseas Development Institute, London.

Romero, M.J. (2015) *What Lies Beneath? A Critical Assessment of PPPs and their Impact on Sustainable Development*. Eurodad, Brussels.

Rostow, W.W. (1959) 'The stages of economic growth', *The Economic History Review* 12(1) 1–16.

Sachdeva, S. (2017) 'RSE scheme set to expand despite oversight concerns', https://www.newsroom.co.nz/2017/07/05/37520/rse-scheme-set-to-expand-despite-oversight-concerns# (accessed 6 April 2018).

Sachs, J.D. (2005) *The End of Poverty: Economic Possibilities of Our Time*. Penguin, New York.

Sachs, J.D. (2012) 'From Millennium Development Goals to Sustainable Development Goals', *The Lancet* 379(9832) 2206–2211.
Scheyvens, R. and Overton, J. (1995) '"Doing Well Out of Our Doing Good": A geography of New Zealand aid', *Pacific Viewpoint* 36(2) 192–207.
Schultz, T.P. (2002) 'Why governments should invest more to educate girls', *World Development* 30(2) 207–225.
Schumacher, E.F. (1973) *Small is Beautiful: Economics as if People Mattered*. Harper and Row, New York.
Schur, M. (2016) 'Public-private partnership funds: Observations from international experience', *ADB East Asia Working Paper Series* no. 6, Asian Development Bank, Manila.
Sears, C. (2019) 'What counts as foreign aid: Dilemmas and ways forward in measuring China's overseas development flows', *The Professional Geographer* 71(1) 135–144.
Six, C. (2009) 'The rise of postcolonial states as donors: A challenge to the development paradigm?, *Third World Quarterly* 30(6) 1103–1121.
Solow, R.M. (1956) 'A contribution to the theory of economic growth', *Quarterly Journal of Economics* 70(1) 65–94.
Swiss, L. (2016) 'World society and the global foreign aid network', *Sociology of Development* 2(4) 342–374.
Taylor, M. (1987) 'Issues in Fiji's development: economic rationality or aid with dignity', in Taylor, M.J. (ed) *Fiji: Future Imperfect?* Allen and Unwin, Sydney, pp.1–13.
The Economist (2015) 'The 169 commandments: development', *The Economist* 414 (8931) 28 March 2015.
The Economist (2017) 'Doing good and doing well: A growing share of aid is spent by private firms, not charities', *The Economist* 423 (9039) 6 May 2017.
Todaro, M.P. and Smith S.C. (2015) *Economic Development*. Pearson Education, Harlow.
Tomlinson, B. (ed.) (2012) *Aid and the Private Sector: Catalysing Poverty Reduction and Development? Reality of Aid 2012 Report*. IBON International, Quezon City.
Toye, J., Harrigan, J. and Mosley, P. (2013) *Aid and Power - Vol 1: The World Bank and Policy Based Lending*. Routledge, London.
Trudeau, J. (2018) 'Prime Minister Justin Trudeau's Address to the 72th Session of the United Nations General Assembly', https://pm.gc.ca/eng/news/2017/09/21/prime-minister-justin-trudeaus-address-72th-session-united-nations-general-assembly (accessed 2 March 2018).
Truman, H.S. (1949) 'Inaugural address of Harry S. Truman', https://avalon.law.yale.edu/20th_century/truman.asp (accessed 9 March 2018).
Trump, D.J. (2018) 'President Donald J. Trump's State of the Union Address', https://www.whitehouse.gov/briefings-statements/president-donald-j-trumps-state-union-address/ (accessed 2 March 2018).
UNDP (2009) *Human Development Report 2009. Overcoming Barriers: Human Mobility and Development*. UNDP, New York.
United Nations (2015a) *Addis Ababa Action Agenda of the Third International Conference on Financing for Development (Addis Ababa Action Agenda)*. United Nations, New York.
United Nations (2015b) *The Millennium Development Goals Report 2015*. United Nations, New York.
Unterhalter, E. (2012) 'Poverty, education, gender and the Millennium Development Goals: Reflections on boundaries and intersectionality', *Theory and Research in Education* 10(3) 253–274.

van Apeldoorn, B., de Graaff, N. and Overbeek, H. (2012) 'The reconfiguration of the global state–capital nexus', *Globalizations* 9(4) 471–486.

Vestergaard, J. and Wade, R.H. (2014) 'Still in the woods: Gridlock in the IMF and the World Bank puts multilateralism at risk', *Global Policy* 6 1–12.

Villarino, E. (2011) 'Top USAID private sector partners: A primer', DevEx https://www.devex.com/news/top-usaid-private-sector-partners-a-primer-75832 (accessed 5 September 2017).

Wade, R. (2010) 'Is the globalization consensus dead?', *Antipode* 41(s1) 142–165.

Waldman, M. (2008) *Falling Short: Aid Effectiveness in Afghanistan*. ACBAR Advocacy Series, Agency Coordinating Body for Afghan Relief (ACBAR), Kabul.

Wallace, T., Bornstein, L. and Chapman, J. (2007) *The Aid Chain: Coercion and Commitment in Development NGOs*, Intermediate Technology Publications/ Practical Action Publications, Rugby.

Williams, G. (1983) 'Boomerang aid', *New Internationalist* 126, https://newint.org/features/1983/08/01/boomerang (accessed 23 November 2019).

Williams, D. (2012) *International Development and Global Politics: History, Theory and Practice*. Routledge, London.

Wohlgemuth, L. (2006) 'Changing aid modalities in Tanzania', *Policy Management Brief* No. 17. European Centre for Development Policy Management, Maastricht.

Woods, N. (2008) 'Whose aid? Whose influence? China, emerging donors and the silent revolution in development assistance', *International Affairs* 84(6) 1205–1221.

Wrighton, N. (2010) 'So what's the problem? International development arrivals in Tuvalu', *Just Change* 18 9.

Wrighton, N. and Overton, J. (2012) 'Coping with participation in small island states: The case of Tuvalu', *Development in Practice* 22(2) 244–255.

Wydick, B., Glewwe, P. and Rutledge, L. (2013) 'Does international child sponsorship work? A six-country study of impacts on adult life outcomes', *Journal of Political Economy* 121(2) 393–436.

Young-Powell, A. (2017) 'Debating the rules: What in-house refugee costs count as aid?', Devex https://www.devex.com/news/debating-the-rules-what-in-house-refugee-costs-count-as-aid-90602 (accessed 12 September 2019).

Index

Note: Page numbers in **bold** refer to figures, page numbers in *italic* refer to tables.

For Product Safety Concerns and Information please contact our EU representative GPSR@taylorandfrancis.com
Taylor & Francis Verlag GmbH, Kaufingerstraße 24, 80331 München, Germany

www.ingramcontent.com/pod-product-compliance
Ingram Content Group UK Ltd.
Pitfield, Milton Keynes, MK11 3LW, UK
UKHW021032180425
457613UK00021B/1141

* 9 7 8 0 3 6 7 4 1 4 8 4 9 *